Powering Autonomous Sensors

María Teresa Penella-López ·
Manuel Gasulla-Forner

Powering Autonomous Sensors

An Integral Approach with Focus on Solar and RF Energy Harvesting

Springer

María Teresa Penella-López
Electrical Engineering
Univ. Politecnica de Catalunya (UPC)
Isabena 15
Monzon, Huesca, 22400
Spain
mayte.penella@gmail.com

Manuel Gasulla-Forner
Electronic Engineering
Univ. Politecnica de Catalunya (UPC)
c/ Esteve Terradas 7
Castelldefels, Barcelona, 08860
Spain
manel.gasulla@upc.edu

ISBN 978-94-007-1572-1 e-ISBN 978-94-007-1573-8
DOI 10.1007/978-94-007-1573-8
Springer Dordrecht Heidelberg London New York

Library of Congress Control Number: 2011929348

Cover design: VTeX UAB, Lithuania

Printed on acid-free paper

Springer is part of Springer Science+Business Media (www.springer.com)

A mi familia

María Teresa Penella

A la meva família

Manuel Gasulla

Preface

Autonomous sensors do not use cables either for transmitting data or for powering electronics. They can be found in wireless sensor networks (WSNs) or in remote acquisition systems. Primary batteries provide a simple design for the powering of autonomous sensors. However, batteries must be replaced when depleted, which can lead to unacceptable maintenance costs whenever the number of autonomous sensors is very large or their accessibility is difficult or impractical. An alternative is to extract energy from the ambient, which is usually known as *energy harvesting*. However, the reduced dimensions of most autonomous sensors lead to a low available power from the energy transducer. Thus, efficient methods and circuits to manage and gather the energy are a must.

This book tackles the powering of autonomous sensors, providing an integral approach by considering both primary batteries and energy harvesting. Two rather different forms of energy harvesting are further dealt with: optical (solar) and radiofrequency (RF). Optical energy presents high energy density, especially outdoors, whereas RF remote powering can be the most feasible option for autonomous sensors embedded into the soil or within structures. Throughout the different chapters, devices such as primary and secondary batteries, supercapacitors, and the energy transducers are extensively reviewed. Then, circuits and methods found in the literature to efficiently extract and gather the energy are presented. Finally, new proposals based in our own research are analyzed and tested. Each chapter is written to be rather independent, incorporating the related literature references into each one.

Chapter 1 first presents autonomous sensors and the feasible energy sources. Both primary batteries and energy harvesters are considered and basic concepts to take into account are detailed. The chapter also provides a general overview of the main problems that appear when designing the power supply of autonomous sensors and refers to some of the most important previous works on this topic.

Chapter 2 analyzes the power consumption of an autonomous sensor and adds some remarks on voltage regulators. In particular, a review is presented, from the point of view of power consumption, on some of the typical devices found in autonomous sensors such as microcontrollers and transceivers. A short introduction to sensors and their signal conditioners is also presented. Finally, a general model based on a pulsed current sink is proposed and experimentally validated.

Chapter 3 reviews the ambient energy sources amenable to powering autonomous sensors: radiant, mechanical, thermal, magnetic, and biochemical. The associated energy transducers are also described and some of the most relevant literature is referenced.

Chapter 4 presents the electrical characteristics and models of primary and secondary batteries, and supercapacitors. Low capacity batteries (up to 3 Ah), both primary and secondary, were experimentally characterized and the parameters of a simplified Randles model are fitted. Several types of supercapacitors were also characterized, obtaining the equivalent series resistances. One of them was further tested to obtain its leakage resistance. Temperature tests, both for batteries and supercapacitors were also carried out. On the other hand, the use of hybrid storage units is proposed in order to reduce the power waste and voltage drop caused by high-resistance batteries. Design criteria for choosing a suitable value of the accompanying supercapacitor are presented. Runtime extensions between 16% and 33% are reported.

Chapter 5 considers optical energy for powering autonomous sensors. A generic model for optical energy transducers, namely solar cells or photovoltaic (PV) panels, is employed to compute the *I–V* and *P–V* curves as well as their dependency on optical power and on temperature. Based on this model, the efficiency of direct-coupled solutions was also computed, ranging from 70% to 90%. With respect to maximum power point trackers (MPPTs), a new open-circuit voltage (OCV) method has been proposed and tested, providing a tracking efficiency greater than 99.5%, higher than that reported for current implementations of OCV methods. Additionally, we devised and tested a novel MPPT method that provides a tracking efficiency greater than 99.6% and an overall efficiency greater than 92% for a PV panel power greater than 100 mW.

Chapter 6 is dedicated to remote RF powering of autonomous sensors and in particular to RF energy harvesters. The RF energy transducer (antenna), matching networks, and ensuing rectifiers are presented. Based on circuits proposed in the literature, extensive simulations for several incoming power levels at the antenna (from -10 dBm to 10 dBm) were performed. We determine that circuit efficiency slightly depends on the number of stages used for the voltage rectifier multiplier, but varies widely with the received power (ranging, when using a shunt-inductor matching network, from 10% at -10 dBm to 80% at 10 dBm). Additionally, as the power level increases, so does the output voltage corresponding to maximum efficiency. For low power levels, *LC* matching networks provided higher efficiencies than shunt-inductor networks, at the cost of greater sensitivity to output voltage variations and to the value of the inductor. Experimental tests were performed with a folded dipole antenna (about 300 Ω), shunt inductor matching, a three-stage rectifier, and a storage unit composed of two series connected NiMH batteries.

María Teresa Penella-López
Manuel Gasulla-Forner

Acknowledgements

This book is the result of the research carried out by the authors during the last five years at the ISI group of the Universitat Politècnica de Catalunya, BarcelonaTech. First, we would like to thank all the technical and economical support from our research group. In particular we want to thank our colleague Oscar López for his inestimable help during the design, implementation and testing of solar maximum power point trackers. His knowledge in power electronics and his enthusiasm have been very inspiring. *Gracias a todos. Gràcies a tots.*

Part of the work on RF energy harvesting was developed during a short stay of Maria Teresa Penella at EPFL, in Lausanne (Switzerland). She would like to thank Michel Declerq, Catherine Dehollain, Nicolas Pillin and Norbert Joehl. *Merci de votre aide.*

We would also like to acknowledge all the institutions and people that supported economically this research. Maria Teresa Penella enjoyed first an FI grant from AGAUR (*Agència de Gestió d'Ajuts Universitaris i de Recerca*) of the Catalan government and afterwards an FPU grant (FPU AP2005-2508) from the Spanish Ministry of education during most of the part of her research. This research has been also funded in part by the Spanish Ministry of Science and Innovation through projects TEC2004-05520 and TEC2007-66331/MIC and by the European Regional Development Union, and by the Spanish-Tunisian cooperation project (AECI-A/01218207). Maria Teresa Penella has been working during the last year at Urbiotica. She would like to thank them the support during the final stage of this research.

Finally, we would like to thank all the people that in one way or another made this book possible: colleagues, friends, and family. Without them this book would be incomplete.

"The last thing one knows in constructing a work is what to put first"

Blaise Pascal (1623–1662) French mathematician and physicist.

Contents

Chapter 1
Introduction

Wireless sensor networks (WSNs) have recently become feasible thanks to reductions in the power consumption of the sensor nodes and to the advent of specific communication standards and commercial transceivers. However, for WSNs to thrive, several challenges must first be addressed, including power supply to the nodes.

In this book we deal with different power supply options for low-power autonomous sensors, concretely, for the nodes of WSNs. Nonetheless, the methods and circuits that we proposed and analyzed, and the conclusions that we reached, can be applied to any load with low-power consumption (*i.e.* less than 1 W), and principally, with a low operating duty cycle.

This chapter briefly describes autonomous sensors and the options for their power supply. It provides a general overview of the main problems that appear when designing autonomous sensor's power supplies and refers to some of the most important previous work on this topic. It also outlines the main challenges to be overcome.

1.1 Autonomous Sensors

Autonomous sensors are measurement devices in which cables are not used for transmitting data or for powering electronics. They serve as the nodes of distributed data acquisition systems and WSNs and are used in numerous fields, including healthcare [1], aerospace [2], and environmental monitoring [3]. To increase their autonomy, autonomous sensors seek to reduce their average power consumption by working in low power modes whenever possible. They spend most of the time in *sleep* (standby) mode, and only *wake up* to perform specific actions—namely, measurement, processing and transmission/reception of data [4].

Figure 1.1 shows a block diagram of a generic autonomous sensor. Sensors convert a signal from a physical or chemical quantity to a corresponding signal in the electrical domain. The sensors to be used depend on the desired application. A signal conditioner (or analog processor) matches a sensor output to the digital processor (usually, a low cost microcontroller). Commercial transceivers are used for wireless

M.T. Penella-López, M. Gasulla-Forner, *Powering Autonomous Sensors*,
DOI 10.1007/978-94-007-1573-8_1, © Springer Science+Business Media B.V. 2011

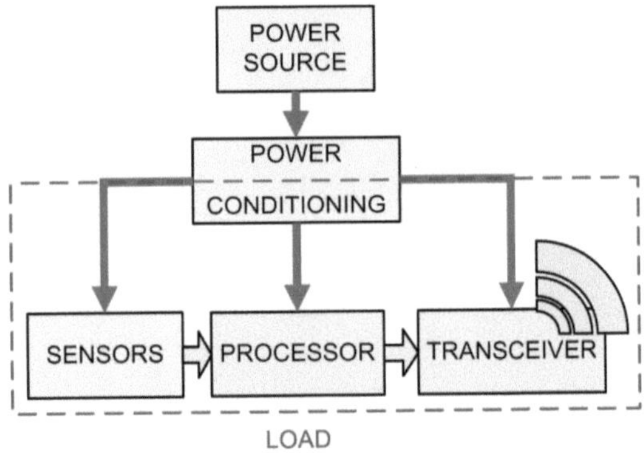

Fig. 1.1 Block diagram of an autonomous sensor

communication. They transmit in the free-licensed ISM bands and can use a proprietary or standard (*e.g.* IEEE 802.15.4) protocol. Power must be provided to the different stages, hereafter denoted as the *load* (sensors, processor and transceiver).

The power source in Fig. 1.1 provides the energy needed by the load. It can be either a primary battery storing a limited amount of energy or an energy-harvesting system. The *power conditioning stage*, whenever necessary, matches the power source to the power supply requirements of the load. Some commercial sensor nodes include the power conditioning stage and accept a wide range of supply voltages. In some designs, the power conditioning stage can be avoided in order to save energy. Conversely, wise use of this stage can reduce the energy demanded from the power source and extend the lifetime of the autonomous sensor. The power source and the power conditioning stage (whenever used) together constitute the *power supply subsystem*. The power supply of the autonomous sensor must provide both the total energy demanded by the load during the expected lifetime of the autonomous sensor and the instant power demanded by the load.

The load can be modeled as a pulsed current sink, as it generally stays inactive (in order to save power) and wakes up from time to time (whether on a schedule or in response to an event). Power consumption in the active mode can be several orders of magnitude higher than in the sleep mode. Therefore, when designing and sizing the power source, the average power demanded by the load ($P_{L,av}$) must be considered. The load's constituent parts must be analyzed in order to obtain a comprehensive model and compute $P_{L,av}$. Although load design falls outside the scope of this work, we do describe some sensors and commercially available microcontrollers and transceivers in Chap. 2—concretely, in terms of their power needs.

1.2 Power Sources for Autonomous Sensors

We considered two types of power sources:

- **Primary batteries**: these provide a simple solution, yet have several drawbacks, including size, limited energy storage, poor indication of remaining capacity, and inefficient extraction of the stored energy.

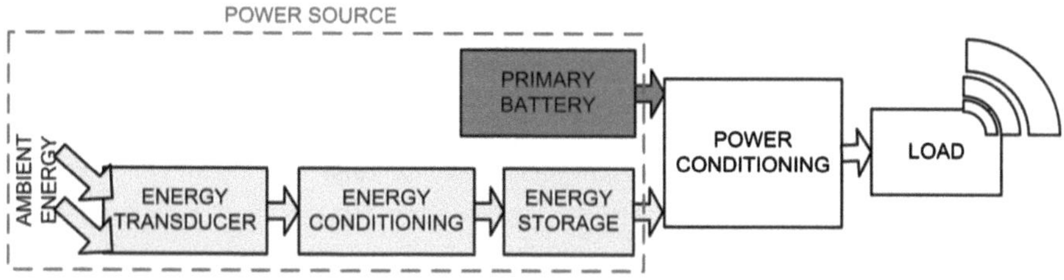

Fig. 1.2 Block diagram of the power supply of an autonomous sensor. Either a primary battery (*dark grey box*) or an energy harvester subsystem (*light grey blocks*) can be used

- **Energy harvesting**: this enables autonomous sensors to run in nearly perpetual operation with little or no maintenance; however, it increases circuit complexity and is a fledgling technology that still poses certain challenges. Optical [5], mechanical [6], thermal [7] or even radiofrequency (RF) energy [8] have been proposed. In this book we focus on optical and RF energy sources.

When choosing the best power source for a given autonomous sensor, a critical analysis of all the power source options must be performed. For example, if the size or weight of the power source is determinant, then the available power density can be used as a decision criterion. In batteries, which contain limited energy, the power density will decrease as the target lifetime of the autonomous sensor increases. Contrariwise, ambient energy sources can provide nearly constant average power density. Thus, the required lifetime of an autonomous sensor will determine the best option [9]. Among ambient energy sources, the specific environmental conditions at the sensor's location will dictate the most suitable option. For example, for autonomous sensors located outdoors, solar energy can be the most feasible solution, whereas for those embedded in the soil or within a structure, remote RF powering may be the sole option. Size, cost, circuit complexity and component count may also be decision criteria. In some cases there will be more than one feasible solution.

Figure 1.2 shows a generic block diagram of an autonomous sensor power supply. The left hand shows the power supply options: either a primary battery (dark grey box) or an energy harvesting system (light grey blocks). Energy harvesting implies more complexity. The (optional) power conditioning stage provides the appropriate power supply to the load.

In order to avoid having to replace primary batteries, batteries with an adequate amount of energy (E_{bat}) for powering the autonomous sensor over its lifetime must be used. Thus, the required E_{bat} for a certain operation lifetime period, T_{op}, can be computed as:

$$E_{bat} = P_{c,av} \cdot T_{op}, \tag{1.1}$$

whereby $P_{c,av}$ is the average power demanded from the power source and includes $P_{L,av}$ plus the average power losses of the different parts, such as that of the power conditioning block.

The deliverable power of a given battery is limited mainly by its series resistance. This resistance in turn depends on several factors, including the battery size and

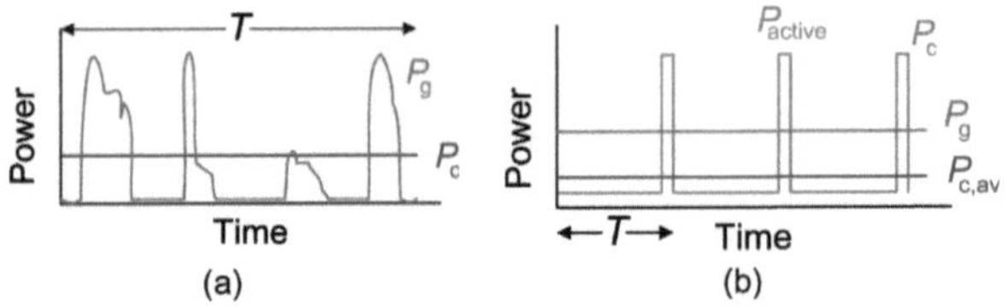

Fig. 1.3 The type of storage required is dictated by the environmental power scenario. For example, (**a**) long-term storage: P_g comes from a variable ambient source (*e.g.* the sun); and (**b**) short-term storage: P_g comes from a constant ambient source, but $P_{active} > P_g$

chemistry. Therefore, the chosen battery must be able to deliver the power demanded by the autonomous sensor during its active time (P_{active}). Alternatively, low power batteries can be used whenever a suitable (super)capacitor is employed in parallel to help provide the required power at the active time [10].

The energy harvester comprises a transducer, an energy conditioning stage and an energy storage unit (Fig. 1.2). The transducer converts ambient energy into electrical energy with a given efficiency. Typical ambient energy sources operate with a certain degree of variability (*e.g.* night-to-day variability when harvesting solar energy), thus an energy storage unit is needed to provide the demanded power flow to the load. An energy conditioning block is used to efficiently charge the storage unit from the output of the transducer.

To maintain sustainable operation of the autonomous sensor when harvesting energy, the average generated power ($P_{g,av}$) must comply with the following:

$$P_{g,av} \geq P_{c,av}. \tag{1.2}$$

Here, $P_{c,av}$ includes $P_{L,av}$, the power losses of the energy and power conditioning blocks and the leakage of the storage unit. $P_{g,av}$ depends on the ambient conditions at the physical location of the node and on the efficiency of the transducer. The storage unit must supplement the power transferred to the load whenever $P_g < P_c$ (*e.g.* Fig. 1.3(a)), whereby P_g and P_c correspond to the instantaneous generated power and consumed power, respectively. Then, for any arbitrary long period of time, T, a long-term storage ($E_{storage}$) unit must fulfill the condition:

$$E_{storage} \geq \max\left\{\int_T (P_c - P_g) \cdot dt\right\}. \tag{1.3}$$

Even if P_g is constant—for example, when the harvested energy comes from permanent indoor lights or from a continuously vibrating machine—a short-term storage unit is needed to withstand the impulsive-type consumption profile of the load. Figure 1.3(b) illustrates this situation, in which $P_{active} > P_g$.

Supercapacitors and secondary (rechargeable) batteries can be used as energy storage units. Supercapacitors offer lower internal impedance and longer lifetimes (in terms of the number of charging/discharging cycles). Moreover, they enable measurement of the remaining energy. However, they have much lower energy density and higher self-discharge, and undergo more severe changes in output voltage in function of extracted or provided charge. Furthermore, they have a much higher

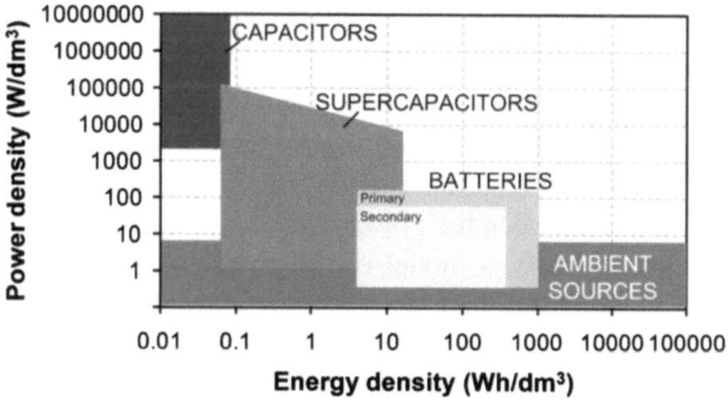

Fig. 1.4 *Power density* versus *energy density* (Ragone plot) of energy sources and storage units. Data obtained from [13] and [11]

cost per energy unit [11]. Secondary batteries and supercapacitors are sometimes used in tandem because of their complementarity [12].

There are several types of secondary batteries. These vary according to characteristics such as voltage, energy density, power density, charging/discharging capabilities, and life (*i.e.* cycle life and calendar life, or *lifetime*) [14].

Figure 1.4 shows a generic graph of the power density versus the energy density of energy sources and storage units. As observed in the figure, power density decreases with increasing energy density. Thus, when used together, elements featuring a higher energy density and elements having a higher power density can be mutually beneficial.

1.3 Challenges

Energy harvesting for autonomous sensors is currently a very hot topic in the research community. However, the work is rather fragmented according to energy source (*e.g.* solar, mechanical and thermal); indeed, an integral approach that includes the different parts of the power supply subsystem of the autonomous sensor (*i.e.* transducer, energy conditioning, storage and power conditioning) is still lacking. Furthermore, primary batteries are generally considered inappropriate power supplies for autonomous sensors, despite the fact that many current autonomous sensors rely on them.

In this book we describe an integral approach to the power supply of autonomous sensors that encompasses both primary batteries and energy harvesters. As for energy harvesters, we have focused on optical and (deliberately radiated) RF energy: the former provides high energy density, whereas the latter provides an alternative when other energy sources (*e.g.* optical, thermal and mechanical) are unavailable.

To achieve an energy-efficient design, all the parts of the power supply subsystem must be studied, optimized and properly interconnected. Designing a power supply subsystem entails three steps. Firstly, power supply requirements must be determined. In autonomous sensors, circuits only wake up when necessary; therefore, the time between wake up periods—and consequently, the energy consumed—strongly depend on the application. Secondly, an appropriate power source must be selected. Primary batteries have limited available energy but are simple to use. Contrariwise, energy harvesting offers unlimited energy over time, but is operationally more complicated. Regardless, the power density is limited by the size of the energy transducer and depends on the physical location of the autonomous sensor. In the third and final step, an energy-efficient circuit must be designed and then implemented in between the power source and the load.

Consequently, a model of the power consumption of the load is first required. This model has to include the power consumption in the different power modes, the time the load remains in each power mode, and time the load takes to change from one state to the other. Additionally, suitable battery models and analyses to evaluate battery performance in autonomous sensors are not present in the literature. The use of supercapacitors in autonomous sensors is also an obscure topic, either as the sole energy storage unit or in combination with batteries in order to provide the required peak power.

On the other hand, achieving optical energy harvesting requires a clear understanding of the related transducer (solar cells), which entails modeling them. It, thus, demands analyses of how to extract the maximum energy from the solar cells, and proposing strategies (*i.e.* methods and circuits) to this end. Literature methods for low power systems are basically adaptations of classical methods for high power panels. However, there is much room for their improvement (*i.e.* increasing power efficiency). New methods and circuits specially designed for low power solar cells must also be explored. Alternatively, the efficiency of simple (direct-coupled) solutions for powering low-power autonomous sensors has not been fairly and systematically analyzed.

Radiofrequency energy harvesting is a fledgling research topic marked by design challenges stemming from rather low power density (*i.e.* microwatt scale). Several analyses have been done for RFID applications, but only a few broad studies have been proposed that include all the blocks of the RF energy harvester for powering an autonomous sensor. Furthermore, there are no clear design guidelines for RF energy harvesting. Some designers try to add as many stages as possible in order to increase the output voltage at a given input power, but they have not necessarily considered simpler solutions. The main objective in RFID is to maximize output voltage, but this does not necessarily maximize power efficiency.

References

1. M. Zakrzewski, S. Junnila, A. Vehkaoja, H. Kailanto, A.M. Vainio, I. Defee, J. Lekkala, J. Vanhala, J. Hyttinen, Utilization of wireless sensor network for health monitoring in home environment, in *Proceedings of IEEE International Symposium on Industrial Embedded Systems, SIES* (2009), pp. 132–135
2. S. Lintelman, R. Robinson, L. Mingyan, L. Bushnell, R. Poovendran, K. Sampigethaya, Secure wireless collection and distribution of commercial airplane health data. IEEE Aerospace and Electronic Systems Magazine **24**, 14–20 (2009)
3. G. Barrenetxea, F. Ingelrest, G. Schaefer, M. Vetterli, Wireless sensor networks for environmental monitoring: the sensorscope experience, in *Proceedings of IEEE International Zurich Seminar on Communications* (2008), pp. 98–101

4. A. Sinha, A. Chandrakasan, Dynamic power management in wireless sensor networks. IEEE Design & Test of Computers **18**, 62–74 (2001)
5. V. Raghunathan, A. Kansal, J. Hsu, J. Friedman, M. Srivastava, Design considerations for solar energy harvesting wireless embedded systems, in *Proceedings of Fourth International Symposium on Information Processing in Sensor Networks* (2005), pp. 457–462
6. P.D. Mitcheson, E.M. Yeatman, G.K. Rao, A.S. Holmes, T.C. Green, Energy harvesting from human and machine motion for wireless electronic devices. Proceedings of the IEEE **96**, 1457–1486 (2008)
7. S. Dalola, M. Ferrari, V. Ferrari, M. Guizzetti, D. Marioli, A. Taroni, Characterization of thermoelectric modules for powering autonomous sensors. IEEE Trans. Instrumentation and Measurement **58**, 99–107 (2009)
8. T. Le, K. Mayaram, T. Fiez, Efficient far-field radio frequency energy harvesting for passively powered sensor networks. IEEE Journal of Solid-State Circuits **43**, 1287–1302 (2008)
9. M.T. Penella, J. Albesa, M. Gasulla, Powering wireless sensor nodes: Primary batteries versus energy harvesting, in *Proceedings of IEEE International, Instrumentation and Measurement Technology Conference, I2MTC* (2009), pp. 1625–1630
10. R.A. Dougal, S. Liu, R.E. White, Power and life extension of battery-ultracapacitor hybrids. IEEE Trans. Components and Packaging Technologies **25**, 120–131 (2002)
11. S.F.J. Flipsen, Power sources compared: The ultimate truth? Journal of Power Sources **162**, 927–934 (2006)
12. C.E. Holland, J.W. Weidner, R.A. Dougal, R.E. White, Experimental characterization of hybrid power systems under pulse current loads. Journal of Power Sources **109**, 32–37 (2002)
13. D. Linden, T.B. Reddy, *Handbook of Batteries*, 3rd edn. (McGraw-Hill, New York, 2001)
14. Woodbank Communications, Electropaedia. Battery and energy technologies. Available on: http://www.mpoweruk.com/. Accessed 22 January 2010

Chapter 2
Load and Power Conditioning

The parts that form the load of autonomous sensors—sensors, processor and transceiver—plus their power and energy needs, must be considered for a model of the load used in designing the power supply subsystem. The load's power requirements will depend on the quantity to be measured and on the system requirements. For example, ambient temperature is usually a slow-changing quantity that enables small duty cycles. A sensor for presence detection has some parts that are always active and others (*e.g.* transmitter) that are randomly triggered by an event (*e.g.* people walking by).

This chapter begins with analysis of the power consumption of the load's constituent parts. A simple and general model that characterizes the overall load power consumption is then proposed. Measurements of the power consumption of an autonomous sensor were performed to validate the proposed model. The power conditioning stage is also considered here, as some devices that form the load, such as the transceivers, generally contain an internal voltage regulator. The power conditioning stage is used to adapt the electrical characteristics of the power source to that of the load. The different strategies that can be addressed are described and compared.

2.1 Load

Figure 2.1 shows the block diagram of an autonomous sensor. The sensor(s), the processor and the transceiver are considered from hereafter as the *load* to be supplied. This load is usually inactive (in sleep mode) and wakes up to perform some of the following actions: measurement, processing and transmission/reception of data [1].

We analyzed the different parts of the load from the perspective of the power consumption in order to obtain a simple and general model that only provides the relevant information.

To obtain a model of the load, we evaluated the power supply characteristics of several sensors, commercial microcontroller units (MCUs) and transceivers. We also

M.T. Penella-López, M. Gasulla-Forner, *Powering Autonomous Sensors*,
DOI 10.1007/978-94-007-1573-8_2, © Springer Science+Business Media B.V. 2011

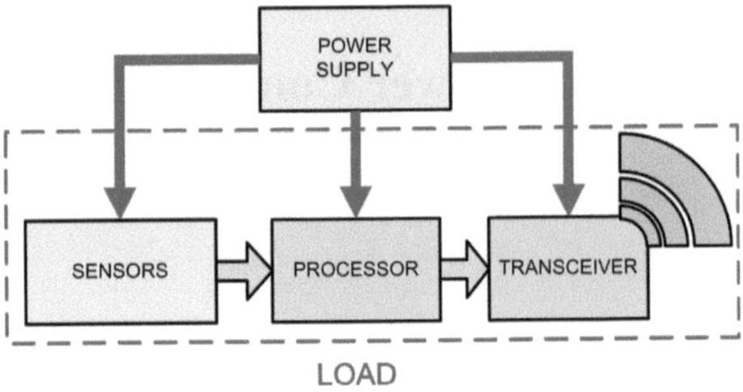

Fig. 2.1 Block diagram of an autonomous sensor

reviewed the relevant literature. However, load design and optimization fall outside of the scope of this thesis.

2.1.1 Sensors and Signal Conditioning

Sensors convert a signal from a physical or chemical quantity into a corresponding signal in the electrical domain. There are many types of sensors, and these vary in their power needs. Nonetheless, regarding the power supply, sensors can be classified as either *modulating* or *self-generating* [2]. In modulating sensors, most of the output signal power comes from the power source. Conversely, in self-generating sensors, output power comes from the input (the quantity to be measured). The output signal of the sensors can be in the millivolt range, and signal-conditioning circuits are used for reaching a suitable signal (units of volts, dc signal) for the analog-to-digital converters (ADCs), which yield a digital code. Signal conditioners can perform any of the following functions: amplification, level shifting, filtering, impedance matching, modulation and demodulation.

Modulating sensors can be resistive, capacitive or inductive. Resistive sensors dissipate energy. Low value resistive sensors are especially prone to high power consumption when used in voltage dividers or bridges. Specific strategies have been proposed to reduce this power consumption [3–5]. A dc voltage or current normally drives the voltage divider or the resistive bridge. For sensors with low resistance, power consumption can be lowered by reducing the driving voltage or current, although at the expense of smaller output and then lower signal to noise ratio. In contrast, there are sensors that require a minimum voltage supply (*e.g.* ultrasonic-based sensors). This voltage supply can be greater than the maximum voltage supply accepted by the other parts of the load; in this case, it would require a dedicated power conditioning stage. The parasitic elements of capacitive and inductive sensors waste some power and additionally these sensors need to be excited with a time-varying voltage or current, which demands specific circuitry that increases power consumption.

Resistive, capacitive and inductive sensors can form part of a variable oscillator that provides an ac signal whose amplitude, period, duty cycle, frequency or phase is proportional to the quantity to be measured. When information is encoded in amplitude, ac signals are typically translated to the dc domain in order to yield a suitable signal for the ADC input. If the phase of the ac signal has to be detected, coherent detection is required. Contrariwise, when the phase carries no information, three basic techniques can be used: peak detection, rms measurement, or mean value measurement after rectification. When information is encoded in either frequency or time, it can be directly measured using an MCU timer. Direct sensor-to-microcontroller techniques [6] have been proposed to directly interface resistive and capacitive sensors to an MCU, which can enable simplification in circuit design and reductions in power consumption. Specific signal conditioners for capacitive sensors built from switched capacitors are also available.

Self-generating sensors provide a voltage or current signal. High amplification is needed for sensors that provide a voltage at the millivolt level (*e.g.* thermocouples). Electrometers, transimpedance amplifiers, and charge amplifiers are used with sensors that provide a relatively high voltage that comes from a high-output impedance source (*e.g.* pH electrodes), or a low current source (*e.g.* piezoelectric sensors) [2]. The operational amplifiers (*op amps*) best suited for building electrometer amplifiers are those whose input stages are based on FET or MOSFET transistors, because these present higher input impedances. Regardless, op amps with low noise, offsets and drifts are required for good resolution and accuracy. Unfortunately, there is a tradeoff between the power consumption and the noise and drift characteristics of op amps. Slew rate and output drive capability also influence their power consumption and performance [3]. Finally, resistances in the megaohm range reduce power consumption, although at the expense of greater thermal noise.

Figure 2.2 shows a generic block diagram of the sensor and signal conditioning stages summarized above, with special emphasis on the power consumption of each part.

Low-pass filters are used to reduce noise for sensors or circuits that employ high resistances, low voltages and/or low currents. The lower the cutoff frequency, the lower the output noise, the higher the resolution, and the longer the response time (and consequently, the energy consumed). Placing the filters right at the sensor's output can reduce the power consumption, as the voltage swing at the filter capacitors will be smaller [4].

If the components of the signal conditioning stage are always active, then they must have ultra-low power consumption. The tradeoff between power consumption and the characteristics of the components must be considered during design. Contrariwise, for components that spend most of their time asleep or powered off, higher-performing units can be employed. Thus, the response and start-up times of the electronic parts involved must be considered. These times sometimes dictate the length of time that the electronics must be activated prior to taking a valid measurement, and therefore, directly affect both the active time and the energy consumption of the autonomous sensor. Finally, the bandwidth of the physical signal to be measured determines the minimum sampling frequency, and consequently, when an autonomous sensor (or at least, its parts required for measurement) must be activated.

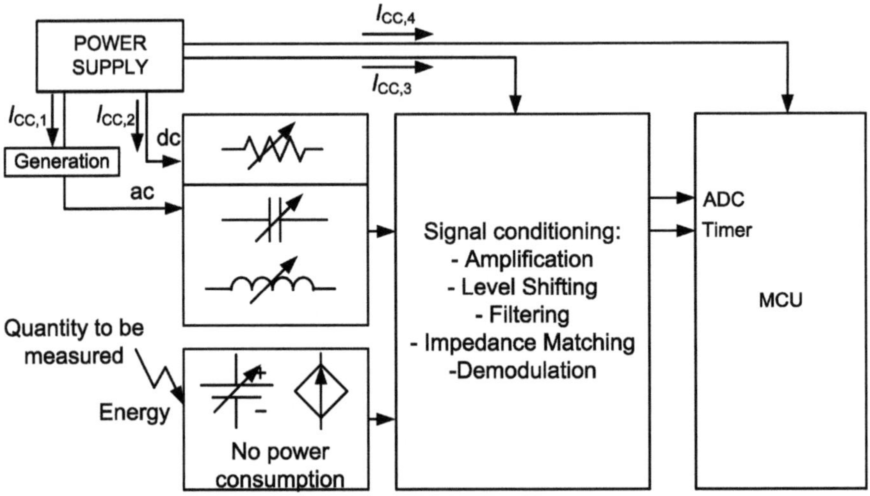

Fig. 2.2 Block diagram of sensors and signal conditioning blocks, illustrating each part's power consumption

2.1.2 Microcontrollers

Low power MCUs are typically used in autonomous sensors to process the digital data and to obtain the measured quantity from a digital code by using additions, multiplications or look-up-tables, among other methods. Most MCUs include ADCs and timers, which enable measurement of dc voltages, frequency or period. Microcontroller units can also implement other functionalities that are not directly related with sensing yet are required for controlling autonomous sensors, including battery protection algorithms, and control of the transceiver or the regulators. Digital signal processors (DSP) are used when high processing capability is needed (*e.g.* for video processing); however, due to their high power consumption, we did not consider them in this work.

Microcontroller units generally lack embedded voltage regulators and accept a wide range of supply voltages (V_{CC}). In order to save power, they are given several power modes (or *states*). There are two main power modes common to all manufacturers: *sleep* and *active*. In sleep mode, all the modules are typically disabled except for a low-power timer (internal or external) to wake up the MCU. In some cases, the MCU can only be woken up by external interruption; this case offers ultra-low power consumption. Data on current consumption (listed on manufacturers' datasheets) in active mode usually accounts for the current consumption of the CPU and the clock; however, if other blocks (*e.g.* ADC, analog comparator, and internal reference) are active, their power consumption must be added to that specified in active mode.

Some tasks, such as writing or erasing flash memory, are power intensive. In addition to active and sleep power modes, manufacturers usually provide several intermediate power modes that may vary widely in current consumption and in

Table 2.1 Power specifications of three low power commercial microcontroller units [7–9], showing typical values

Model	V_{cc} (V)	$I_{MCU,sleep}$ (μA)	$I_{MCU,active}$ (μA)	Wakeup time from sleep	Wakeup time from reset
TI, MSP430x11x2	1.8–3.6 >2.2 @ $f = 4$ MHz > 2.8 @ $f = 8$ MHz	0.1 @ 2.2 V (ram retention)	200 @ 2.2 V; 1 MHz	6 μs @ $f < 3$ MHz	2 ms
Atmel, ATtiny24A	1.8–5.5 >2.7 @ $f > 4$ MHz >4.5 @ $f > 10$ MHz	0.1 @ 1.8 V	210 @ 1.8 V; 1 MHz	750 ns @ $f = 8$ MHz	4 ms
Microchip, PIC16LF72X	1.8–3.6 >2.3 @ $f > 16$ MHz	0.02 @ 1.8 V	110 @ 1.8 V; 1 MHz	7 μs	65 ms

the components that are active. To minimize consumption, the input/output pins must be handled carefully, and all the unnecessary parts should be disabled. Table 2.1 shows the current consumption in sleep ($I_{MCU,sleep}$) and active ($I_{MCU,active}$) modes at 25 °C for three similar commercial MCUs. As observed in the table, a difference of three to four orders of magnitude can be expected from $I_{MCU,sleep}$ to $I_{MCU,active}$.

Figure 2.3 shows how $I_{MCU,active}$ increases with increasing voltage supply (Fig. 2.3(a)) and operating frequency (Fig. 2.3(b)). These plots are based on an Atmel MCU [8], but data for devices from other manufacturers are similar [7]. As observed in Fig. 2.3(a), $I_{MCU,active}$ increases linearly with increasing supply voltage. Consequently, the dissipated power depends on the square of the supply voltage [10]. Thus, working at the minimum possible supply voltage is recommended. Alternatively, temperature appears to only slightly affect $I_{MCU,active}$. Figure 2.3(b) shows that $I_{MCU,active}$ also increases linearly with increasing oscillator frequency (f). However, as the frequency increases, the time needed for executing instructions decreases, leading to similar overall energy consumption for the MCU. Nonetheless, working at the lowest supply voltages in order to save power is only possible at the lowest clock frequencies ($f < 4$ MHz for $V_{CC} = 1.8$ V in Fig. 2.3(b)).

Dynamic voltage scaling (DVS) [1] is a power-saving technique that instantaneously adapts the supply voltage to the needs of the autonomous sensor. For example, when the autonomous sensor runs at maximum frequency, the supply voltage is set at a high value, whereas when the autonomous sensor sleeps, the supply voltage is reduced to its minimum value. This technique exploits all the available resources at a minimum energy cost, but leads to more complex circuits and control.

In sleep mode, the MCU works with a low-power timer at a fixed frequency; consequently, $I_{MCU,sleep}$ is not frequency dependent, but does vary with temperature and voltage (Fig. 2.4). As observed in the figure, the variation of $I_{MCU,sleep}$ with supply voltage is not linear under these conditions, in which temperature has a greater effect. The $I_{MCU,sleep}$ in this figure is roughly three orders of magnitude lower than the $I_{MCU,active}$ in Fig. 2.3(b).

When switching between power modes, time is spent on waking up the components or on storing the MCU current state. Therefore, power (and consequently,

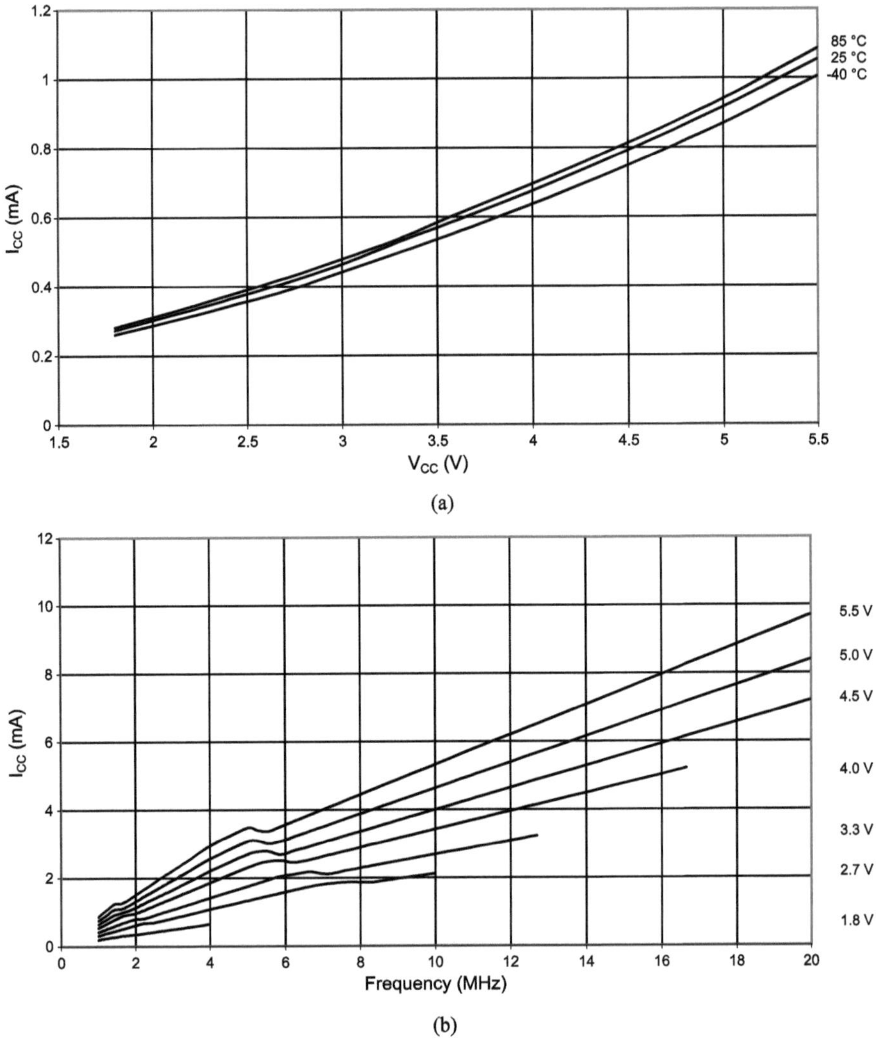

Fig. 2.3 $I_{\mathrm{MCU,active}}$ versus (**a**) supply voltage at several temperatures and 1 MHz, and (**b**) frequency at several supply voltages. Data from the ATtiny24A Atmel microcontroller unit [8]

energy) is wasted during this time. Usually, as the power consumption of a state decreases, the time required to wake up from this state increases, resulting in non-negligible energy consumption [11]. Moreover, every component that can be switched on and off has a stabilization time that must be accounted for in every specific application. Thus, if the amount of energy saved by a component while sleeping is similar to that wasted in its stabilization time, thus no energy will be saved. For this scenario, the component should probably be kept awake.

Table 2.1 presents the wake up times from the sleep state and from reset for each MCU. As observed in the Table, the values differ by model; thus, these times can

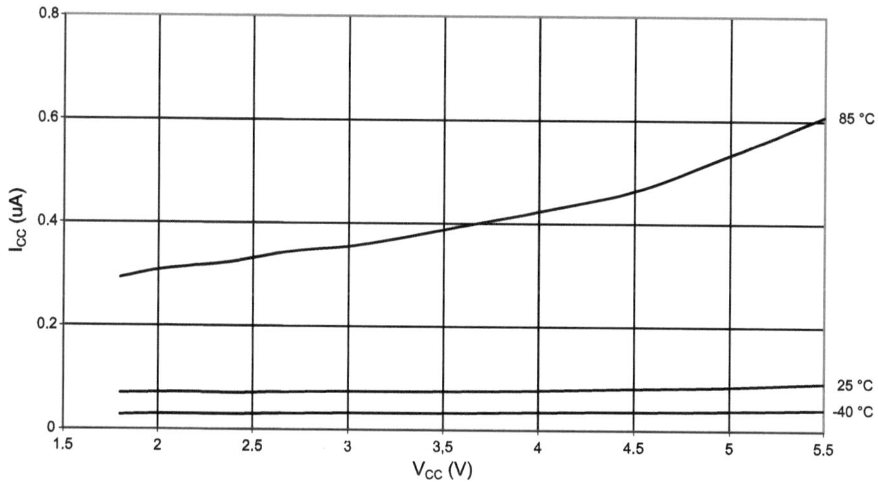

Fig. 2.4 Variation of $I_{MCU,sleep}$ with voltage supply and temperature for the ATtiny24A Atmel microcontroller unit [8]

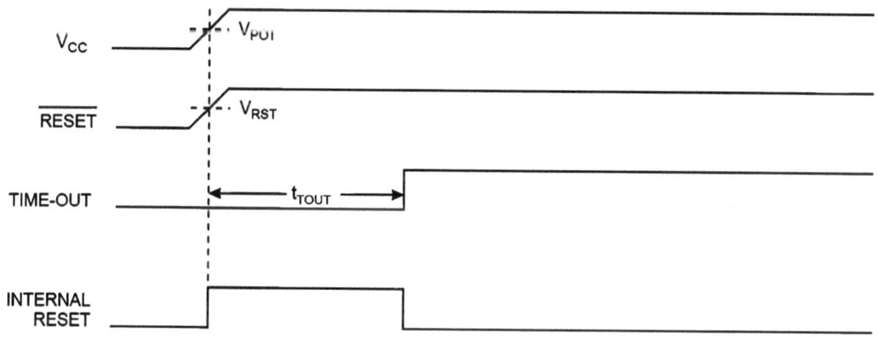

Fig. 2.5 Power on reset timing for the ATtiny24A Atmel microcontroller unit [8]

be used as criteria for selecting the MCU that offers the lowest power consumption for a given autonomous sensor. However, these data must be interpreted with caution, as manufacturers differ in their specifications and characterization methods.

Waking up from reset takes much longer than waking up from sleep. Figure 2.5 shows that each time an MCU is powered up from its off state (no supply voltage), a reset (*power on reset [POR]*) is performed to generate a suitable supply voltage. Therefore, the time that an MCU takes to be completely operational from reset must be considered when powering the node for the first time or whenever the power supply is removed between measurements (t_{TOUT} in Fig. 2.5).

2.1.3 *Transceiver*

A transceiver transmits and receives data wirelessly via RF waves, according to a proprietary or standard protocol. The Zigbee specification [12] is optimized for networks with a very low data rate and low power consumption. Bluetooth low-energy technology, recently included in Bluetooth version 4.0, offers minimal power consumption at high data rates (1 Mbit/s) [13].

This section only covers ZigBee transceivers, as Bluetooth low-energy technology was not available during the development of this work. It begins with a brief summary of the remarkable characteristics of the Zigbee specification, and then analyzes the power supply characteristics of some commercial models.

2.1.3.1 The Zigbee Specification

The Zigbee specification is based on IEEE 802.15.4 standard and defines a series of protocols for efficient, facile communication of multiple devices. It can operate at 2.4 GHz (16 channels), 915 MHz (10 channels, US free band) and 868 MHz (1 channel, Europe free band). The modulation used is direct-sequence spread spectrum (DSSS) and the medium access protocol is Carrier Sense Multiple Access with Collision Avoidance, (CSMA/CA), which directly impacts power consumption, as the channel must be listened to before a packet can be sent. Several network topologies can be implemented, including tree, star or mesh. In most commercial ZigBee transceivers this process is done automatically and is transparent to the user. The network is established as follows: the first ZigBee device present in the environment establishes a network and names itself as parent device; then, as new nodes arrive, they add as children to this network [12]. The ZigBee protocol defines three types of devices:

- Full function devices (FFD): These route messages between ZigBee devices. When they are allocated in large networks, they should remain active as long as possible. Alternatively, if these nodes are allocated in small networks, then they can be periodically woken up, always in synchronization with the nodes they have to route. These devices can also incorporate sensors.
- Reduced-function devices (RFD): These are end devices that periodically poll their parent for any message addressed to them. They are typically the sensor nodes of the network. There are two types:
  - Sleepy end device (SED): These are always joined to the same parent device and are unable to leave the network. They are used for static networks.
  - Mobile sleepy end device (MSD): After being inactive for three seconds, these devices are erased from the network; upon reactivation, they search for a new parent device from the same network or another one.
- Coordinator (COO): This node is the FFD that initiates the network and parent of all other devices in the network. A network can only have one COO.

Table 2.2 Power specifications of ZigBee transceivers [14–17]

Model	V_{cc} (V)	$I_{RF,sleep}$ (μA)	$I_{RF,rx}$ (mA)	$I_{RF,tx}$ (mA)	Wakeup time from sleep	Wakeup time from reset
Atmel, AT86RF230 (system-on-chip)	1.8–3.6	0.02	15.5	16.5 (@ 3 dBm)	880 μs	3 ms
Ember, EM 250 (system-on-chip)	2.1–3.6	1	36	36 (@ 5 dBm)	Not specified	Not specified
TI, CC2520 (transceiver only)	1.8–3.8	<1 @ full V_{CC} range	18.5 @ full V_{CC} range	33.6 (@ 5 dBm and full V_{CC} range)	300 μs	≥ 0.3 ms (clock dependent)
TI, CC2430 (system-on-chip)	2–3.6	0.3	26.7	26.9 (@ 0 dBm)	120 μs	1.7 ms @ 32 kHz

The type of ZigBee device configured will influence the power consumption of the node. Network management overhead (*e.g.* acknowledgements) can also strongly influence power consumption if not carefully designed.

2.1.3.2 Commercial ZigBee Transceivers

The main characteristics of the four marketed ZigBee transceivers that we studied are listed in Table 2.2. Of these devices, three are system-on-chip (SOC) integrated circuits that combine an RF transceiver and an MCU. They offer several power modes, and every part of the system can be independently powered on or off by setting a specific register. All the transceivers have an internal linear low dropout regulator (LDO) to supply a stable voltage for some of the system parts (*e.g.* the radio). Consequently, the current consumption in active states barely changes with voltage supply variations (Fig. 2.6(a)) and is quite higher than that of MCUs (Sect. 2.1.2). The transmit current ($I_{RF,tx}$) is slightly influenced by the transmitted power. Table 2.2 displays this current for the maximum transmitted power allowed, and shows the reception ($I_{RF,rx}$) and sleep ($I_{RF,sleep}$) currents as well. In most cases $I_{RF,tx}$ is four to five orders of magnitude higher than $I_{RF,sleep}$. Current consumptions are specified at $V_{CC} = 3$ V. Wakeup time varies among different commercial transceivers and depends on the power-down mode from which they stem. As with MCUs, this time is not negligible.

Figure 2.6 shows how $I_{RF,tx}$ (a) and $I_{RF,sleep}$ (b) depend on the supply voltage for a CC2520 transceiver [16]. The variation in $I_{RF,tx}$ above 2 V is very low (<5%); thus, $I_{RF,tx}$ can be considered a constant current. As such, the power consumption in this mode is directly proportional to the supply voltage. In contrast, $I_{RF,sleep}$ almost triples when the supply voltage varies by 2 V (in sleep state the LDO is disconnected). Thus, working at the minimum possible supply voltage is desirable for both MCUs and transceivers in order to reduce power consumption.

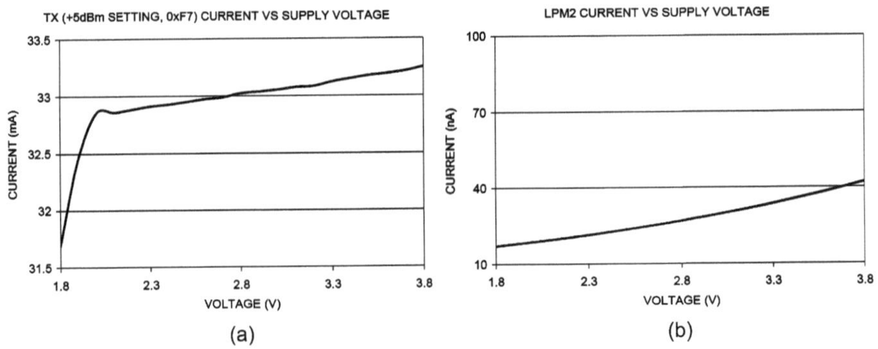

Fig. 2.6 (**a**) $I_{\mathrm{RF,tx}}$ @ +5 dBm versus supply voltage and (**b**) $I_{\mathrm{RF,sleep}}$ versus supply voltage. Data from a CC2520 transceiver [16]. Courtesy Texas Instruments

2.1.4 Load Model

The literature contains various models for wireless sensor nodes or networks. Some researchers have extensively analyzed MAC or network protocols to obtain a Markov Chain that describes the nodes [1, 18] or the network [19] from the perspective of power consumption. These studies are mostly used for simulation of network lifetime. Other researchers have evaluated the sensor node as a single element that does not interact with the rest of the network; they use a load model to test proposed models for other parts of the autonomous sensor, such as a supercapacitor [20] or an energy harvester [21]. In contrast, we sought a simple model that easily describes the load power profile.

Dynamic power management (DPM) is a power-saving strategy based on the connection or disconnection of the parts of a system that are not used, which generates several power consumption states. It has been applied to wireless sensor networks since their onset [1]. Most authors [1, 18, 19] have dealt with four or five active states (depending on the blocks of the load that remain active; see Fig. 2.1).

We considered a model with just three active states, which represent the main functions (transmit, receive and idle), plus one sleep state (Fig. 2.7) [19]; nonetheless, a different number of states can be considered. Because not all the functionalities are always done, not all the active states need to be visited when the load wakes up. In sleep state, all the parts remain inactive, except for some timers or the circuitry required for waking up the node. In idle state, only the sensing parts and the processor parts remain active. In the transmit or receive state, the transceiver is encoding and then sending packets, or receiving and then decoding packets. The MCU and the transceiver are both active at this time; hence, the total power consumption in these states is equal to the sum of the current consumption of both components. The energy spent to switch between states (*e.g.* $E_{\text{sleep-idle}}$ in Fig. 2.7(a)) can be considered by simply extending the time spent in the initial or destination state. The average current consumption ($I_{\mathrm{L,av}}$), assuming a periodic performance of the node, gives:

$$I_{\mathrm{L,av}} = \frac{t_{\mathrm{tx}} \cdot I_{\mathrm{tx}} + t_{\mathrm{rx}} \cdot I_{\mathrm{rx}} + t_{\mathrm{idle}} \cdot I_{\mathrm{idle}} + t_{\mathrm{off}} \cdot I_{\mathrm{off}}}{T}, \tag{2.1}$$

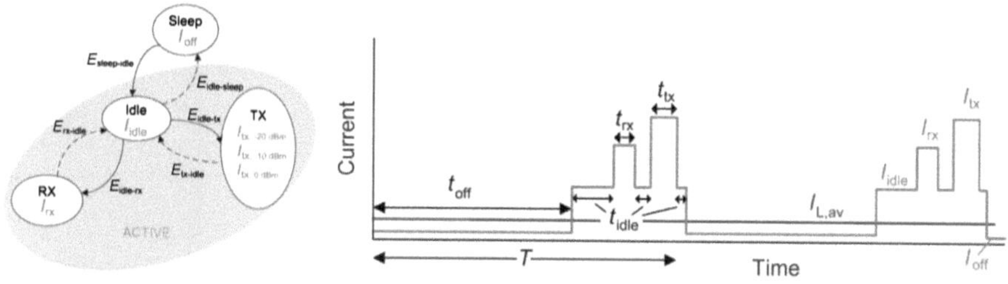

(a) (b)

Fig. 2.7 (**a**) State diagram when considering three active states plus one sleep state. (**b**) A generic current profile of the load. Both adapted from [11]

whereby $T = t_{tx} + t_{rx} + t_{idle} + t_{off}$; t_{tx}, t_{rx}, t_{idle} and t_{off} are the time intervals that the node is transmitting, receiving, in an idle state and sleeping, respectively; and I_{tx}, I_{rx}, I_{idle} and I_{off} are the current consumptions in the corresponding states (Fig. 2.7(b)). A state diagram can be drawn (Fig. 2.7(a)). Autonomous sensors usually wake up to the idle state, and then may transmit or receive. They then return to the idle state prior to sleep. Nonetheless, if no information has to be transmitted (*e.g.* when recording a measurement), the autonomous sensor simply wakes up to the idle state.

As stated in Chap. 1, the average current consumption is necessary for sizing the power supply. As later explained in Chap. 4, the peak value of the current can influence the voltage behavior of the power supply, especially at relatively high values of output resistance.

2.1.4.1 Measurements

To assess our model, we measured the current consumption of an autonomous sensor that used an EM 250 transceiver. We configured the autonomous sensor to send a data packet every 5 min and disabled all the other network default functionalities. The sensor node measured battery voltage and ambient temperature. The sensors and signal conditioning circuits functioned at 1.8 V, so they were powered from the internal regulator of the transceiver, which produced this voltage.

We computed the average current consumption of the sensor node at 3 V by measuring the current consumption when packets are sent and the current consumption when in sleep mode. For these measurements, the sensor node was powered from a dc power supply. Firstly, we used a DSO8104A oscilloscope and a 10 Ω resistor, placed in series with the dc power supply, to measure the time intervals and the current when the sensor node woke up (Fig. 2.8). Despite the fact that we had disabled all the network functionalities, every approximately 32 s the sensor node woke up for just 2 ms. These network functionalities cannot be disregarded. Figure 2.8(b) offers a more detailed illustration of the time at which the packet was sent (every 300 s, which corresponds to $t = 142$ s in Fig. 2.8(a)). As observed in the figure, the sensor node woke up three times during a period of 1.5 s. This phenomenon

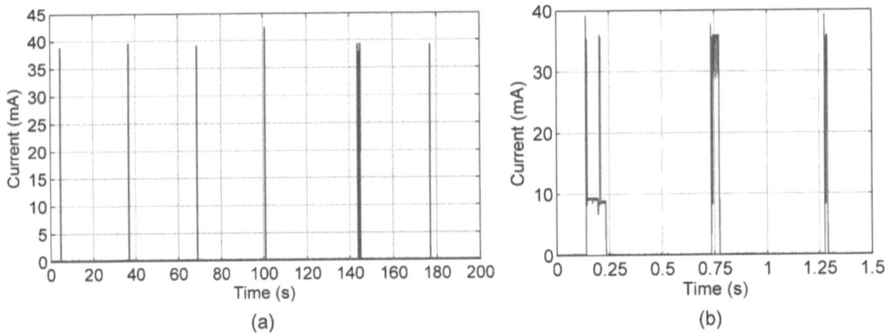

Fig. 2.8 Power consumption of a sleepy end device. (**a**) 200 s acquisition; and (**b**) magnified area of plot at the time when a packet was sent

Table 2.3 Times and currents in every state of the load working as a sleepy end device

State	Time (ms)	Current (mA)	Average current (μA/%)	
Idle	101.6	9.3	3.224	20.68
Transmit (data + ACK)	32.7	36	3.924	25.17
Transmit (network functionalities)	17.1	36	2.052	13.16
Receive (data + ACK)	13.7	34.9	1.594	10.22
Sleep	299831.8	0.0048	4.797	30.77
TOTAL	300000	–	15.59	100

of waking up more than once each time a packet had been sent was due to polling the parent for acknowledgement (ACK) [22]. The current behavior in Fig. 2.8(b) is similar to that described in Fig. 2.7.

We next measured the current while the node was sleeping, using the 34401A (Agilent) multimeter in the 10 mA scale: the value was 4.8 μA at 3 V. This current consumption was mostly due to a resistor divider used to measure the supply voltage. Table 2.3 summarizes the times and currents spent in every state of the load every 5 min. We distinguished between network functionalities and data packets in order to determine if the ZigBee network functionalities consumed a significant amount of power. We were able to easily compute the value of $I_{L,av}$ by using (2.1): the result was 15.59 μA. We also computed the average current of each state. For this T (5 min), network functionalities corresponded to 13% of $I_{L,av}$. The greatest contribution was from the sleep state. The network functionalities and the sleep state consume an even greater share of power in applications with a larger T. Thus, in these cases, minimizing the power consumption in sleep state and disabling all unnecessary network communications are essential.

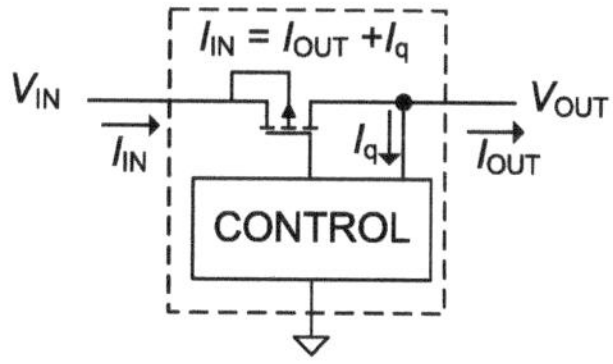

Fig. 2.9 Linear regulator using a MOSFET

2.2 Power Conditioning

Power conditioning is used to adapt the power delivered by the power source to the load's power needs (Fig. 1.2). The power source can be either a primary battery or an energy harvester, which nearly always includes a storage unit (*i.e.* a secondary battery or supercapacitor). The supply voltage needs of autonomous sensors are described earlier in this chapter and the electrical characteristics of primary batteries and storage units are described in Chap. 4.

Power conditioning is mainly achieved through the use of voltage regulators. Three main types can be outlined [23]: *linear regulators*, *switching regulators*, and *charge pumps* (or *non-inductive switching regulators*). Regulators can be skipped whenever the voltage range of the power source fits within that of the load. However, as we have seen, the power consumption of the load is reduced by working at the minimum supply voltage. When using a regulator to achieve this, its power waste should be less than the power gain. Whenever the load or some of its parts go to sleep, the corresponding voltage regulator should be disabled or put into shutdown in order to save power. Regardless, designers must assure that the MCU (or the circuit that controls the power conditioning stage) never loses its own power supply—otherwise, the load will not be able to restart.

In a voltage regulator, efficiency (η) is defined as the ratio of power output ($P_{OUT} = V_{OUT} I_{OUT}$) to power input ($P_{IN} = V_{IN} I_{IN}$), as observed in (2.2), where V and I refer respectively to the voltage and current.

$$\eta = \frac{P_{OUT}}{P_{IN}} = \frac{V_{OUT} I_{OUT}}{V_{IN} I_{IN}}. \tag{2.2}$$

In the remaining of this section we describe the main features of the different types of voltage regulators as well as a control strategy used in switching regulators for low power loads.

2.2.1 *Linear Regulators*

Linear regulators are based on an active device, such as a bipolar junction transistor or field effect transistor, operating in its *linear region* (Fig. 2.9), or a passive device (*e.g.* a Zener diode), operating in its *breakdown region*. The regulating device is employed as a variable resistor: it continuously adjusts a voltage divider network to maintain a constant output voltage. Linear regulators provide a regulated output that is lower than their input ($V_{OUT} < V_{IN}$). A dropout between input and output is

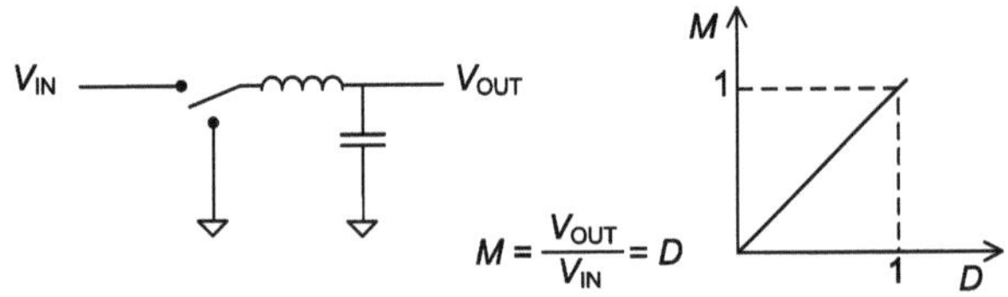

Fig. 2.10 A buck converter and its conversion ratio

required, although its value can be very small (this is the case for *low-dropout linear regulators*) [24].

I_{IN} equals I_{OUT} plus the current consumed by the control circuitry (I_q). Equation (2.3) assumes $I_q \ll I_{OUT}$. Efficiency increases as the voltage difference between input and output (ΔV) is decreased. Maximal efficiency is achieved when the difference between input and output is equal to the dropout voltage (V_{DO} in (2.3)).

$$\eta = \frac{V_{OUT}}{V_{IN}} = \frac{1}{1 + \frac{\Delta V}{V_{OUT}}},$$
$$\eta_{MAX} = \frac{1}{1 + \frac{V_{DO}}{V_{OUT}}}. \tag{2.3}$$

Commercial IC linear regulators are easy to use. For example, regulators with fixed output voltage require only two external capacitors to function. Furthermore, they do not suffer from any significant noise, since they do not require any switching elements, and therefore, their circuit board layout is not critical. The power that is not delivered to the load is dissipated as heat in the regulator. Thus, when used in high-power applications these regulators demand board space for thermal dissipation [25].

2.2.2 *Switching Regulators*

Switching regulators can *step up* (increase), *step down* (decrease), or *step up or down* (increase or decrease) their input voltage. Their name derives from the fact that they use a switch in combination with a filter to change the voltage level. The conversion ratio ($M = V_{OUT}/V_{IN}$) is the ratio of output voltage to input voltage. It depends on the switch's duty cycle (D). The most common topologies for switching regulators are *buck converter* (which step down; see Fig. 2.10), *boost converters* (which step up; see Fig. 2.11) and *buck-boost converter* (which step up or down, depending on D, and invert V_{IN}; see Fig. 2.12) [26]. Switches are usually implemented with transistors and diodes. Synchronous converters use two transistors that are alternatively activated, which increases the efficiency. The duty cycle is usually controlled in order to maintain a fixed output voltage. Section 2.2.4 describes a control strategy suited for low-power loads.

Contrariwise to linear regulators, ideal efficiency is 100%, as no dissipative elements are present. In practice, though, internal power losses (P_{LOSS}), which are classified as either *dynamic* or *static*, limit the maximum efficiency. Dynamic losses

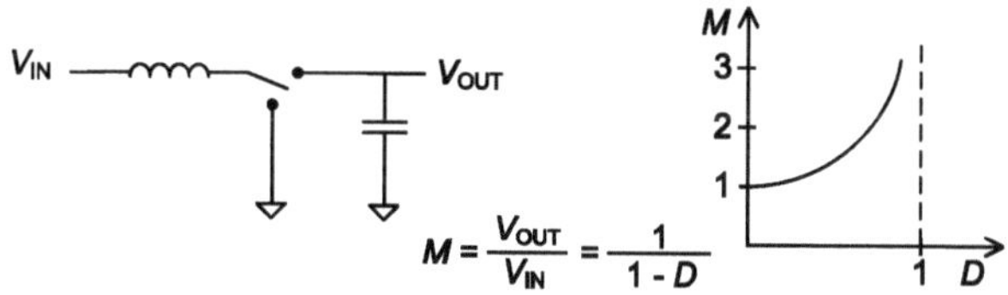

Fig. 2.11 A boost converter and its conversion ratio

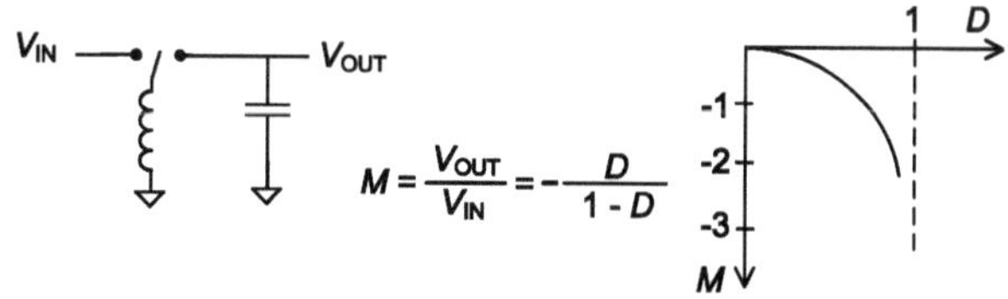

Fig. 2.12 A buck-boost converter and its conversion ratio

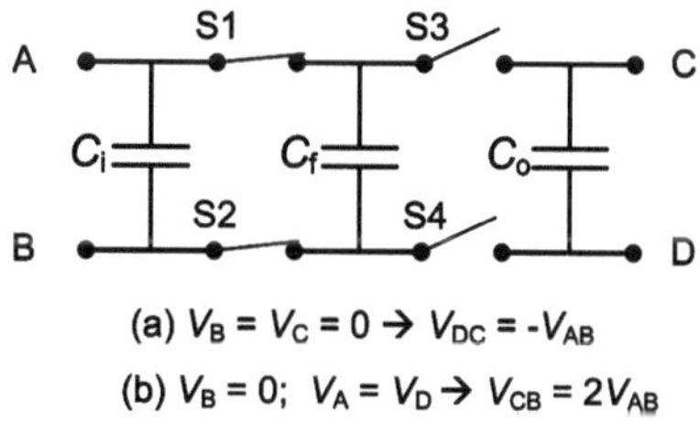

Fig. 2.13 (**a**) An inverter charge pump and (**b**) a doubler charge pump

are load dependent (*i.e.* they increase as the load current increases) and derive from parasitic elements, the conduction losses of the switch, and the switching losses (which also depend on switching frequency). In contrast, static losses are constant, regardless of the load current. They are associated to the control part of the converter (*i.e.* the quiescent current going into the control circuit). Dynamic losses predominate at higher I_{OUT}, whereas static losses predominate at lower I_{OUT}. Obviously, the lower the P_{LOSS}, the greater the efficiency. Efficiency for milliampere load currents is typically greater than 90%, but decreases for lower I_{OUT}. Nonetheless, several power saving techniques have been developed to improve efficiency at low output currents. Switching regulators generate electromagnetic interference (EMI) from the switched current of the inductor. Thus, in order to lessen the effects of EMI on the analog parts of the entire system, the board should be carefully laid out. Electromagnetic interference can also be minimized through the use of filters.

2.2.3 Charge Pumps

Charge pumps are switching converters that periodically switch capacitors to step down, step up, or step up or down voltages; they do not employ inductors for these functions. They are based on the charge-transfer process. Figure 2.13 shows the charge-transfer basic circuit. When S1 and S2 are closed the charge is transferred from the input capacitor (C_i) to the *flying capacitor* (C_f). In contrast, when S1

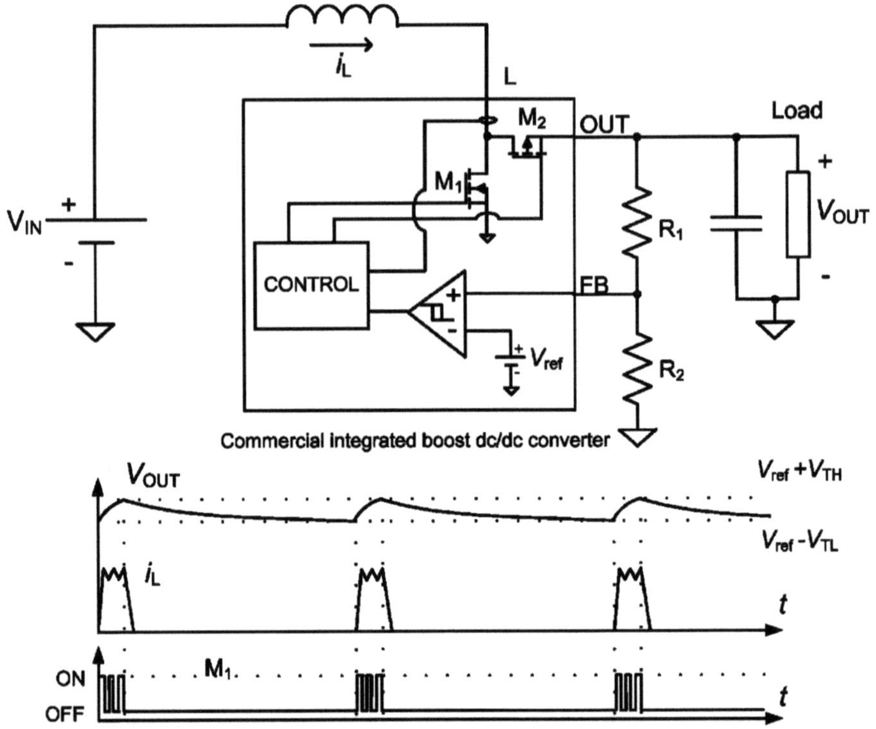

Fig. 2.14 Pulse frequency modulation used in a boost regulator at light loads [29]

and S2 are subsequently opened, and S3 and S4 are closed, there is charge transfer between C_f and the output capacitor (C_o). Assuming no losses, in steady state $V_A - V_B = V_C - V_D$. If B and C are connected and grounded, then an *inverting charge pump* is obtained (Fig. 2.13(a)). Alternatively, if A and D are connected, and B is grounded, then a *doubler charge pump* (*i.e.* one which doubles the input voltage) is obtained (Fig. 2.13(b)) [27].

Analogously to conventional switching regulators, different control strategies for the switches are used in order to maintain a fixed output voltage. Further, they also present dynamic and static power losses. Commercial devices feature several power-saving modes to improve efficiency at low load currents. Charge pumps generate some noise, but it is smaller than that of conventional switched regulators. These devices support loads in the range of hundreds of milliamperes.

2.2.4 Control Strategy

Switched regulators and charge pumps require a control strategy for turning their switches on and off. *Pulse width modulation* (*PWM*) is the most widely used control strategy for high load currents [26]. Alternatively, *pulse frequency modulation*

(*PFM*), which is based on hysteresis control of the output voltage, is a control strategy employed to achieve low power consumption by the converter and high efficiency with light loads [28]. We only dealt with PFM, as it is intended for low-power loads.

Figure 2.14 shows a schematic of an inductor-based synchronous PFM boost regulator together with the temporal evolution of the output voltage (V_{OUT}), the inductor current (i_L), and the control signal of the internal switch M_1. Switches M_1 and M_2 remain off until V_{OUT} falls below a low threshold value ($V_{ref} - V_{TL}$). Then, the switches alternately turn on in order to transfer electrical charge from the input to the output until V_{OUT} reaches a high threshold value ($V_{ref} + V_{TH}$), at which point the process restarts. For light loads, the inactive period increases, maintaining low overall power consumption and high efficiency. The same concept can be applied to buck or buck-boost switching regulators and to charge pumps. Several regulators employing this strategy are commercially available.

2.3 Conclusions

This chapter describes the power consumption performance of all the parts of autonomous sensors: sensors, signal conditioners, MCUs and transceivers.

Regarding the power supply, the sensors can be classified as either *modulating* or *self-generating*. Modulating sensors modulate a signal and can be resistive, inductive or capacitive. The power dissipated by resistive sensors can be sizeable for some low value sensors. Ideally, inductive and capacitive sensors do not consume power. However, both their parasitic elements and the required ac excitation circuitry waste power. Self-generating sensors generate an electrical signal (sometimes quite small) from the physical quantity to be measured. A signal conditioning stage is usually required to interface sensors to the MCU (ADCs or timers). Specially, those sensors providing small-level signals may need precision op amps that inherently present high-power consumption. As a result, special care must be put on the design of sensors and their signal conditioning in order to minimize the power consumption of this stage. Several techniques proposed in the literature can be used to reduce the power consumption of some sensors. Finally, designers must select devices that meet the system requirements with the lowest power consumption.

Microcontroller units acquire the signal from the sensors and perform other tasks. Their power consumption in active mode is two to three orders of magnitude less than that of the transceiver. However, if not properly handled, the active blocks inside the MCU and the input/output ports can considerably increase the final power consumption of the MCU. Additionally, the power consumption can vary several orders of magnitude between active and sleep states. So, MCUs must be carefully handled in order to keep their power consumption in a low level.

The transceiver communicates with the other nodes of the network via an RF signal. It is usually considered the most power hungry part, but some sensors can surpass its power consumption, especially if not correctly designed. The power consumption during data transmission is four orders of magnitude higher than the power consumption during sleep.

The MCU and the transceiver consume more current at increasing supply voltage. Thus, one power-saving strategy is to operate at the minimum supply voltage (provided that the required functionalities can run at this voltage level). However, this may not always be desirable: for example, an MCU cannot operate at maximum frequency at its minimum supply voltage. Hence, power-saving techniques such as DVS can also be considered.

Autonomous sensor operation is based on DPM, whereby all the parts that are not used are disabled, resulting in several power consumption modes. We accounted for these when developing a model for the autonomous sensor that considers the load as a pulsed current sink. Power consumption varies widely among these modes, and in very low duty cycle applications the relative power consumption during sleep can be crucial. The described load model is a tool for sizing the power supply.

The voltage supply from the power source has a voltage range that often falls outside of that accepted by the load. Thus, a power conditioning stage is required. Furthermore, as the power consumption of the autonomous sensor decreases with decreasing supply voltage, power conditioning can be considered in all cases for increasing lifetime. A thorough study of efficiency and circuit complexity (among other criteria) must be considered when designing the power conditioning stage.

We have briefly presented here several power conditioning circuits: *linear regulators*, *switching regulators* and *charge pumps*. Linear regulators can only decrease voltage. They are simple and generate no significant noise. Switching regulators and charge pumps can increase or decrease their input voltage and require a control strategy to turn their switches on and off. Both power-conditioning circuits suffer from *dynamic* and *static* power losses that ultimately determine their efficiency at a given load. Charge pumps do not use inductors, and consequently, enable simpler and more compact designs.

PFM is a control strategy used in commercial switching regulators and charge pumps intended for light loads. It is based on hysteresis control of the output for shortening the active time of the internal components of the regulator (and consequently, its power consumption).

Future work should propose smarter power conditioning circuits and control strategies for their application in very low power autonomous sensors.

References

1. A. Sinha, A. Chandrakasan, Dynamic power management in wireless sensor networks. IEEE Design & Test of Computers **18**, 62–74 (2001)
2. R. Pallas-Areny, *Sensors and Signal Conditioning*, 2nd edn. (New York, 2001)
3. J. Williams, Micropower circuits for signal conditioning. Available on: http://www.linear.com/pc/downloadDocument.do?navId=H0,C1,C1154,C1004,C1139,P1247,D4118. Accessed April 2010

4. M. Conner, Run for your life: Ultra-low power systems designed for the long Haul.EDN, electronic design newsletter (2005, April 2010). Available on: http://www.edn.com/article/471243-Run_for_your_life_Ultralow_power_systems_designed_for_the_long_haul.php
5. R. Casanella, R. Pallas-Areny, On the design of low-power signal conditioners for resistive sensors, in *Proceedings of XIX IMEKO World Congress* (2009), pp. 787–791
6. F. Reverter, R. Pallàs-Areny, *Direct Sensor-to-Microcontroller Interface Circuits* (Marcombo, Barcelona, 2005).
7. Texas Instruments, MSP430x11x2, MSP430x12x2 mixed signal microcontroller. Available on: http://focus.ti.com/lit/ds/symlink/msp430f1122.pdf. Accessed July 9, 2009
8. Atmel, 8-bit AVR microcontroller with 2k/4k bytes in-system programmable flash, ATtiny 24A, ATtiny44A. Available on: http://www.atmel.com/dyn/resources/prod_documents/8183S.pdf. Accessed July 2008
9. Microchip, PIC16F72X/PIC16LF72X datasheet. 28/40/44-Pin flash-based, 8-bit CMOS microcontrollers with nanoWatt XLP technology. Available on: http://ww1.microchip.com/downloads/en/DeviceDoc/41341C.pdf. Accessed 2009
10. L. Benini, G. De Micheli, Introduction, in *Dynamic Power Management. Design Techniques and CAD Tools* (Kluwer Academics, Norwell, 1998), pp. 1–39
11. B. Bougard, F. Catthoor, D.C. Daly, A. Chandrakasan, W. Dehaene, Energy efficiency of the IEEE 802.15.4 standard in dense wireless microsensor networks: modeling and improvement perspectives, in *Proceedings of Design, Automation and Test in Europe*, vol. 1 (2005), pp. 196–201.
12. ZigBee Alliance, ZigBee alliance home page. Available on: http://www.zigbee.org/en/index.asp. Accessed June 2010
13. Bluetooth SIG, Bluetooth low energy technology. Available on: http://www.bluetooth.com/English/Products/Pages/Low_Energy.aspx. Accessed June 14, 2010
14. Atmel, Low power 2.4 GHz transceiver for ZigBee, IEEE 802.15.4, 6loWPAN, RF4CE and ISM applications. Available on: http://www.atmel.com/dyn/resources/prod_documents/doc5131.pdf. Accessed July 2009
15. Ember, EM250 single-chip ZigBee/802.15.4 solution. Available on: http://www.ember.com/pdf/120-0082-000_EM250_Datasheet.pdf. Accessed July 2009
16. Texas Instruments, CC2520 datasheet 2.4 GHz IEEE 802.15.4/ZigBee RF transceiver. Available on: http://focus.ti.com/lit/ds/symlink/cc2520.pdf. Accessed July 9, 2009
17. Texas Instruments, CC2430. A true system-on-chip solution for 2.4 GHz IEEE 802.15.4/ZigBee. Available on: http://focus.ti.com/lit/ds/symlink/cc2430.pdf. Accessed July 9, 2009
18. D. Jung, T. Teixeira, A. Savvides, Sensor node lifetime analysis: Models and tools. ACM Trans. Sensor Networks **5**, 1–33 (2009)
19. C.-F. Chiasserini, M. Garetto, An analytical model for wireless sensor networks with sleeping nodes. IEEE Trans. Mobile Computing **5**, 1706–1718 (2006)
20. G.V. Merrett, A.S. Weddell, A.P. Lewis, N.R. Harris, B.M. Al-Hashimi, N.M. White, An empirical energy model for supercapacitor powered wireless sensor nodes, in *Proceedings of 17th International Conference on Computer Communications and Networks, ICCCN* (2008), pp. 1–6
21. A.F. Susu, A. Acquaviva, D. Atienza, G. De Micheli, Stochastic modelling and analysis for environmentally powered wireless sensor nodes, in *Proceedings of 6th International Symposium on Modeling and Optimization in Mobile, Ad Hoc, and Wireless Networks and Workshops, WiOPT* (2008), pp. 125–134
22. Telegesis, TG-APP-ETRX2Power 100. ETRX2 wireless mesh networking module. Application note—power consumption. Available on: http://www.telegesis.com/downloads/general/TG-APP-ETRX2Power-100.pdf. Accessed June 25, 2009
23. P. Rako, Creating the power subsystem in modern design. EDN, electronic design newsletter (2006, June 2010). Available on: http://www.edn-europe.com/creatingthepowersubsysteminmoderndesign+article+402+Europe.html
24. Dallas Semiconductor Maxim, Linear regulators in portable applications. Available on: http://pdfserv.maxim-ic.com/en/an/AN751.pdf. Accessed May 15, 2007

25. Dallas Semiconductor Maxim, Choosing the right power-supply IC for your application. Available on: http://pdfserv.maxim-ic.com/en/an/AN737.pdf. Accessed May 15, 2007
26. R.W. Erickson, D. Maksimovic, *Fundamentals of Power Electronics*, 2nd edn. (Kluwer Academic, Norwell, 2001)
27. Maxim. Dallas Semiconductor, Properties of the charge-pump voltage splitter. Available on: http://www.maxim-ic.com/app-notes/index.mvp/id/718. Accessed December 2007
28. B. Sahu, G.A. Rincon-Mora, An accurate, low-voltage, CMOS switching power supply with adaptive on-time pulse-frequency modulation (PFM) control. IEEE Trans. on Circuits and Systems I **54**, 312–321 (2007)
29. O. Lopez-Lapena, M.T. Penella, M. Gasulla, A new MPPT method for low-power solar energy harvesting. IEEE Trans. on Industrial Electronics **57**, 3129–3138 (2010)

Chapter 3
Ambient Energy Sources

For decades solar and wind energy have been exploited for diverse applications. The past decade has seen major research effort dedicated to diversifying energy sources for powering autonomous sensors. Thomas et al. [1] extensively described many energy sources for powering small-scale unmanned systems. Based on their initial classification, we have categorized harvested energy for powering autonomous sensors into five types: *radiant, mechanical, thermal, magnetic*, and *biochemical* (Fig. 3.1). Here, we consider energy already present in the environment, although in some cases energy can be expressly distributed, in a process called *remote powering*. Other analyses and typical power densities of energy sources can be found in the literature; for example, in [2] or [3]. The choice of the ambient energy source strongly depends on the ambient energy density at the autonomous sensor's location and on the load's power needs.

Energy transducers must be used to convert the ambient energy source into an electrical output (Fig. 3.1). Efficiency here is defined as the ratio of the transducer's output power to its input power. This chapter chiefly describes the above related energy sources together with the corresponding energy transducers. Given that this book deals with solar and radiofrequency (RF) energy harvesting, solar and RF transducers, together with their associated energy conditioning stages, are further treated in Chaps. 5 ("Optical Energy Harvesting") and 6 ("Radiofrequency Energy Harvesting").

3.1 Radiant Energy

Radiant energy is emitted as electromagnetic waves, which includes radioactive energy. Radioactive materials have extremely high energy densities, and their half-lives span from fractions of a second to millions of years. Radioactive properties can be found in uranium, nickel or phosphorous isotopes. However, because these materials can be extremely dangerous, they must be confined to a closed volume. The energy from radioactive sources that emit alpha and beta particles can be converted by piezoelectric or betavoltaic elements, which determine their efficiency [4].

M.T. Penella-López, M. Gasulla-Forner, *Powering Autonomous Sensors*,
DOI 10.1007/978-94-007-1573-8_3,

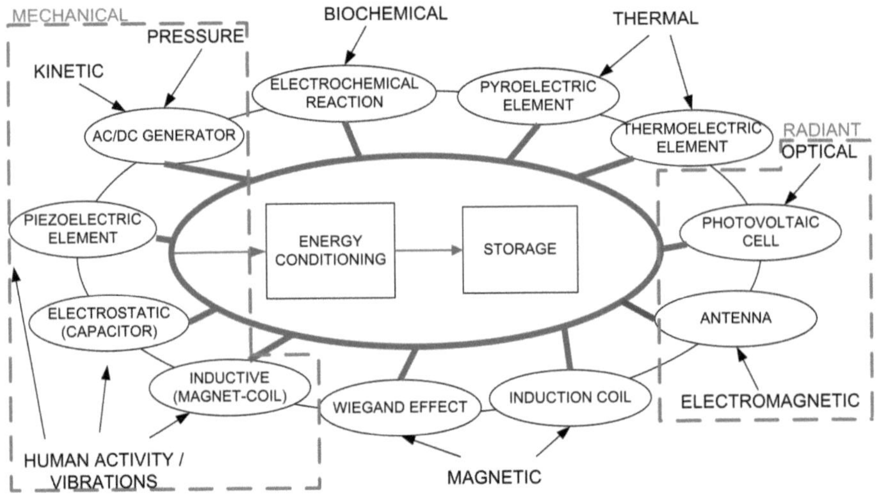

Fig. 3.1 Energy sources and their respective transducers for powering autonomous sensors. Adapted from [1], with additional power sources added

3.1.1 Optical Energy

Optical energy [2, 5], also referred as solar energy, accounts to the portion of the electromagnetic spectrum spanning infrared energy to ultraviolet light.

Outdoors, optical energy chiefly comes from the sun. Part of the energy radiated by the sun is absorbed by the atmosphere; the rest arrives to the surface of the Earth as both direct and indirect radiation. On cloudy days, indirect radiation predominates. The available power density depends on several factors, including latitude, weather conditions, the day of the year, and the time of day. For example, at noon on a sunny day, the incident light on the surface of the earth has a power density of roughly 100 mW/cm^2 (or 1000 W/m^2). Outdoors, solar cells must be aimed at the sun to maximize energy harvesting. There are published tables that provide the recommended tilt from the horizontal plane for fixed solar panels at a given latitude [6].

Indoor optical energy comes from outdoors (*e.g.* near windows) and from artificial illumination (*e.g.* incandescent and fluorescent bulbs). Power density in a room depends on the distance to the source; the size, shape, and spectral density of the source; and the spatial distribution of light in the room. The factors affecting spatial light distribution comprise surface reflection, windows absorption and transmission, indoor and outdoor obstacles, light sources distribution, or lighting recommendations [5]. The hourly availability of these sources depends on room use. Power density, when the lights of a room are on, typically ranges from 100 $\mu W/cm^2$ to 1000 $\mu W/cm^2$.

Optical energy is harvested via photovoltaic (solar) cells. A solar cell is basically a semiconductor diode with a large p-n junction in the plane of the cell. A simple model includes a dc current source with a diode in parallel. Most current solar cells

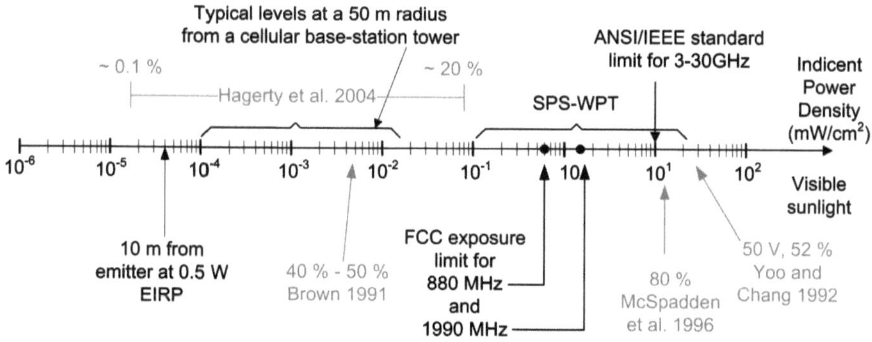

Fig. 3.2 The range of ambient radiofrequency power available, showing efficiencies reported by various researchers. Adapted from [7]. SPS-WPT: satellite power system—wireless power transmission

are silicon-based, and fall into one of three categories: monocrystalline, polycrystalline, and amorphous. Their efficiencies at full sun range from as high as 30% for laboratory-tested monocrystalline cells down to 5% for commercial amorphous cells [1]. Crystalline cells are better suited for outdoor use, whereas amorphous cells work better indoors.

3.1.2 Radiofrequency Energy

Radiofrequency (RF) signals are deliberately radiated by broadcasting station antennas and cellular phone antennas. The preferred antennas for harvesting power from ambient RF sources are either broadband and circularly polarized [7], because in most cases, there is either a broad range of frequencies and different polarizations or there is lack of detailed information on the ambient RF waves. Except for at locations near transmitters, the available ambient power density (S) is in the nanowatt range (Fig. 3.2), which is too low to be useful. Figure 3.2 shows typical values for available ambient RF power as well as efficiencies reported by other research groups. However, near transmitters the power is much higher.

Energy can also be deliberately radiated without license on the industrial, scientific and medical (ISM) applications frequency bands. In this case, S can be obtained from the following equation:

$$S = P_{\mathrm{EIRP}} \frac{1}{4\pi d^2}, \tag{3.1}$$

whereby P_{EIRP} is the effective radiated power, which is related to that of an isotropic radiator (an ideal antenna), and d is the distance from the transmitter. This formula is for free-space propagation; for other scenarios, the power decays more rapidly with distance, unless the location features a structure that acts as a waveguide (*e.g.* in corridors). Inside a building there are other factors that affect S, including multipath propagation and reflections; thus, calculating the real S for an indoor location requires accurate simulations or field measurements. In Europe, the maximum P_{EIRP} for ISM frequency bands is regulated by European Communications Office (ECO) Recommendation ERC/REC 70-03 (see Annex 1 in [8]). For example, at the 868 MHz band the maximum P_{EIRP} allowed is 0.5 W. The available power from the antenna (P_{AV}) depends on S and on the effective aperture (area) of the antenna (A_{e}), as shown in (3.2), below (known as the *Friis relation*):

$$P_{\text{AV}} = S \cdot A_{\text{e}} = S\frac{\lambda_{\text{RF}}^2}{4\pi}G_{\text{r}} = P_{\text{EIRP}}G_{\text{r}}\frac{\lambda_{\text{RF}}^2}{(4\pi d)^2}, \tag{3.2}$$

whereby $A_{\text{e}} = \lambda_{\text{RF}}^2 G_{\text{r}}/4\pi$; G_{r} is the antenna gain; and λ_{RF} is the wavelength. Equation (3.2) clearly demonstrates that the higher the frequency (λ_{RF} is inversely proportional to frequency), the lower the powering range. In Europe, the most popular frequency bands for ISM applications are centered at 433 MHz, 868 MHz and 2.4 GHz. For this book, we chose 868 MHz (see Chap. 6) to avoid any interference with communication systems used in WSNs (which usually operate at 2.4 GHz) and because the 433 MHz band would have increased the power range for the same emitted power at the cost of larger antenna dimensions.

An RF transducer (antenna) can be modeled as a voltage ac source in series with an output impedance [9, 10]. Antenna shapes and dimensions vary widely, and several distinct antenna designs have been employed in RF harvesting and radiofrequency identification (RFID) applications.

3.2 Mechanical Energy

Mechanical energy harvesting is a very active research topic that has spawned numerous reviews (*e.g.* [11]). It is based on kinetic energy [12], the sources of which include liquid or gas flow, vibrations, human activity, and pressure variations (*e.g.* acoustic noise and atmospheric pressure). Available power in flow energy increases cubically with an increase in the speed of the liquid or gas. Low-level vibrations occur in machinery, outdoor windows, and transport vehicles; they produce frequencies between 50 Hz and 200 Hz and acceleration amplitudes between 1 m/s^2 and 10 m/s^2. Human activity can actively or passively generate kinetic energy. Active human power requires deliberate movement, whereas passive human power exploits common daily activities (*e.g.* heel strike while walking); indeed, nearly all the energy produced in daily activities stems from when legs are bent and from heel strike.

Mechanical energy can be coupled by one of the following conversion principles [13]: electrostatic, piezoelectric and electromagnetic. Reported efficiencies are 0.32% [11]; 0.5% (for polyvinylidene fluoride [PVDF]) to 20% (for lead zirconate titanate [PZT]) [14]; and 6% [15], respectively. The generated power is ac and must be rectified in order to power autonomous sensors.

Electrostatic converters are based on variable capacitors. One plate is fixed and the other changes with the mechanical force, thereby changing the value of the capacitance. Among these converters' attractive features is that they are IC-compatible: microelectric variable capacitors can be fabricated via silicon micromachining techniques. Energy can be extracted via charge-constrained or voltage-constrained approaches. Charge and voltage are related through capacitance ($Q = C \cdot V$); therefore, a change in capacitance produces variations in either voltage (charge constraint [16]) or current (voltage constraint [17]). However, these converters have several drawbacks: they demand a power source for extracting energy; they require synchronization between the movement and the extraction cycles; and, in the charge constraint scheme, they operate at high voltages, which limits their implementation to more expensive integration processes [18]. The harvested energy can be quantified using the equation below [19]:

$$E = \frac{1}{2}Q^2\left(\frac{1}{C_{\min}} - \frac{1}{C_{\max}}\right), \tag{3.3}$$

whereby Q is the charge and $C_{\min}$ and $C_{\max}$ correspond to the minimum and maximum capacitances achieved during a extraction cycle.

Reversible electric polarization in response to strain from mechanical stress is called the *piezoelectric effect*. Piezoelectric materials are anisotropic: their properties vary according to the direction of force, and the orientation of the polarization and the electrodes. Several operational modes can be employed for piezoelectric harvesters, the most widely used of which are modes 33 and 31. Mode 33 is used to extract energy from impact or when harvesting energy from passive human power. Mode 31 is more suited for cantilever structures with a proof mass at the free end. Cantilevers are usually bimorphic structures comprising two piezoelectric materials bound together, with a shim in between them. For power extraction, cantilevers typically operate at their resonance frequency. Piezoelectric transducers can be electrically modeled as an ac current source in parallel with a capacitor [20]. The main challenge with piezoelectric harvesters is integrating them into an IC. However, in this case, unlike electrostatic generators, they do not require a voltage source.

Electromagnetic induction, based on Faraday's law (see (3.4)), refers to generation of a voltage (v) in a conductor (typically, a coil) located within a magnetic flux ϕ, usually generated by a permanent magnet. The generated voltage depends on various factors, including the strength of the magnetic field, the velocity of the relative motion between the coil and the magnet, and the number of turns in the coil. Electromagnetic energy transducers can be electrically modeled as low-level ac voltage sources with low series impedance.

$$v = -\frac{d\phi}{dt}. \tag{3.4}$$

However, since magnets are quite difficult to integrate into circuits, integrated prototypes of electromagnetic energy transducers designed this way show very low output powers. Some commercially available energy harvesters (*e.g.* the ones from Perpetuum Ldt) are based on this principle.

Figure 3.3 shows a schematic of each type of mechanical energy transducer.

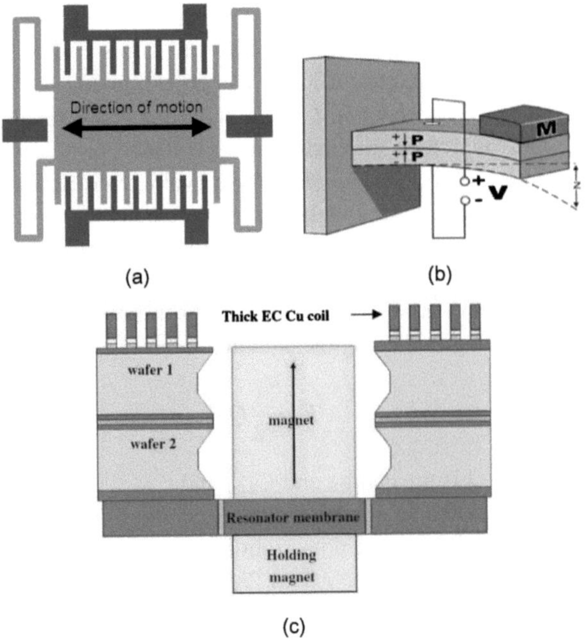

Fig. 3.3 Examples of (**a**) an electrostatic harvester[1] [21]; (**b**) a bimorphic piezoelectric cantilever[1] [21]; and (**c**) an integrated electromagnetic energy harvester[2] [22]

3.3 Thermal Energy

Thermal energy can be converted to electrical energy by thermoelectric or pyroelectric transducers, among others. The former rely on spatial variations in temperature, whereas the latter rely on temporal variations in temperature.

Pyroelectric converters, like piezoelectric transducers, can be modeled as an ac current source in parallel with a capacitor. The pyroelectric current $i_P(t)$ varies with the rate of change of the pyroelectric material temperature, as described by:

$$i_P(t) = p'A\frac{dT}{dt}, \tag{3.5}$$

whereby p' is the component of the pyroelectric coefficient vector $\boldsymbol{p}$ that is orthogonal to the electrode surface of area, A; and dT/dt denotes the temperature change over time. Pyroelectric currents of less than 1 μA for temperature changes of roughly 50 K and 73 K have been reported for a 16 cm^2 PZT converter [23] and a 1.44 cm^2 PZT converter, respectively [24]. More recently, Guyomar et al. [25] employed a

[1]Reprinted from Computer Communications, vol. 26. Roundy, S., Wright, P.K., Rabaey, J., A study of low level vibrations as a power source for wireless sensor nodes, pp. 1131–1144. Copyright (2003), with permission from Elsevier.

[2]With kind permission from Springer Science+Business Media: Design and implementation of mechanical resonators for optimized inertial electromagnetic microgenerators, vol. 14, 2008, pp. 653–658, Serre, C., Pérez-Rodríguez, A., Fondevilla, N., Martincic, E., Martínez, S., Morante, J., Montserrat, J. and Esteve, J. Fig. 1. *Microsystem Technologies*.

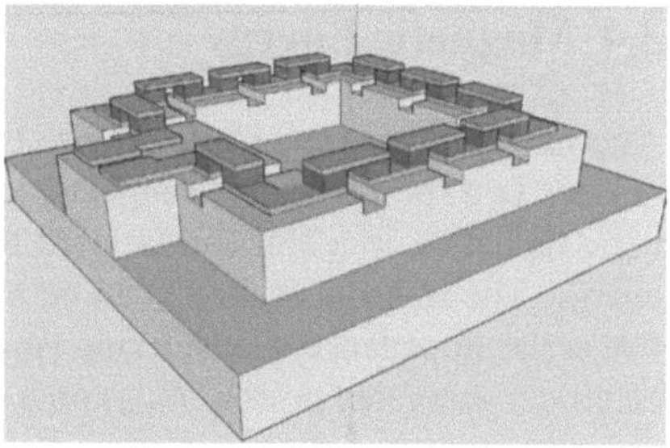

Fig. 3.4 Thermoelectric generator schematic[3] [26]

20 cm^2 PVDF converter to harvest up to 0.32 mW from a sinusoidal temperature variation of 7 K of amplitude and a frequency of 0.2 Hz.

Thermoelectric transducers and generators, which exploit the *Seebeck effect*, comprise several pairs of alternating p- and n-type semiconductor blocks arranged thermally in parallel and connected electrically in series (Fig. 3.4). They can be modeled as a dc voltage source in series with a low-value resistance. The open circuit voltage (V_{OC}) of each thermoelectric element depends on the temperature difference (ΔT) between the hot and cold sides, and on the Seebeck coefficients (α_1 and α_2), which are material dependent (see (3.6), below).

$$V_{oc} = \alpha_1 \Delta T - \alpha_2 \Delta T. \tag{3.6}$$

P- and n-type semiconductors have Seebeck coefficients of opposite sign, which provides maximum voltage. Thermoelectric generators are typically built from Bi_2Te_3 or Poly-SiGe. The thermal resistance of the thermoelectric generator should be matched to that of the thermal source in order to maximize the harvested power. Optimizing the extracted power requires additional elements—namely, a radiator for efficient dissipation of heat into the environment, and structures, known as *thermal shunts*, that direct the heat that passes between the hot and cold plate into the thermocouple legs [26].

When harvesting thermal energy, the maximal efficiency is given by the Carnot efficiency:

$$\eta = (T_H - T_C)/T_H \tag{3.7}$$

whereby T_H and T_C are the hot and cold temperatures (measured in Kelvin), respectively. For example, for a temperature difference of 5 K at an ambient temperature of 293 K (T_C) the Carnot efficiency is 1.6%. The thermal conductivity of silicon is 140 W/mK; thus, the heat flow due to convection in a 1 cm-thick silicon material is 7 W/cm^2. So, the theoretical maximal output power is 117 mW/cm^2 [12]. However, the reported efficiencies are well below the Carnot efficiency. For thermoelectric generators, efficiency can theoretically be up to 17% of Carnot efficiency at body-ambient temperature gradients [27]. Carnot efficiencies of 0.02% to 0.05% have been reported for pyroelectric harvesters [25].

[3]Reprinted from Solid-State Electronics, vol. 53, Vullers, R.J.M., van Schaijk, R., Doms, I., Van Hoof, C., Mertens, R. Micropower energy harvesting, pp. 684–693. Copyright (2009), with permission from Elsevier.

3.4 Magnetic Energy

Alternating current magnetic energy [1] is available near ac current-carrying conductors (*e.g.* power grids), magnets, and electric rotating machinery. Electromagnetic induction in a coil can be used to convert ac magnetic energy into electrical energy. The transducer (coil) can be modeled as an ac voltage source with inductive series impedance. Considering Ampere's law and Faraday's law (see (3.4)), the electrical generated power (P_{rms}) on a coil with impedance Z can be computed as:

$$P_{rms} = \frac{N^2 A_{coil}^2 \mu_{coil}^2 f^2 I_{rms}^2}{Zr^2}, \tag{3.8}$$

whereby N is the number of turns; A_{coil}, the surface of the coil; μ_{coil}, the permeability of the coil core; f and I_{rms}, the frequency and amplitude, respectively, of the ac current source; and r, the distance from the ac current source to the coil. When designing an energy harvester, f and I_{rms} are given; the other parameters can be tuned to harvest the desired power. In contrast to electromagnetic energy harvesters, the magnetic field in this case is not caused by physical movement, but rather by an alternating current. In [1], the authors predicted a generated power of 5.7 mW for a coil with $A_{coil} = 1\ \text{cm}^2$, $N = 1000$ and $\mu_{coil} = 1000$, which is placed at $r = 1$ m from a conductor with $I_{rms} = 10$ A and $f = 60$ Hz. This principle has recently been applied to energy harvesting from a high voltage transmission line [28].

Wireless non-radiative power transfer through strongly coupled resonant objects (also known as *magnetic resonance*) has been reported by researchers at the Massachusetts Institute of Technology [29], who achieved very promising overall efficiencies of up to 40% with a power range of 2 m. Fulton Innovation LLC seems to use that principle in its *ecoupled* technology. The concept is also similar to that already used in low frequency RFID systems in which energy is transferred via magnetic coupling from a reader in order to power nearby tags. On the other hand, inductive powering has been largely applied for biomedical implants [30, 31].

Electromagnetic materials [32] have recently been proposed for harvesting magnetic energy with relatively high efficiency. The *Wiegand effect* [33], which stems from reverse switching of the core magnetization in a ferromagnetic wire upon application of a strong magnetic field, can also produce electrical energy, but affords only very low efficiencies. The electrical energy is produced because the switching of the core induces a voltage in a coil.

3.5 Biochemical Energy

Biochemical energy [34] can be extracted by using fuel cells. In a typical cell, fuel is continuously fed to the anode compartment and an oxidant (*e.g.* oxygen from air) is continuously fed to the cathode compartment; electrochemical reactions occur at the electrodes, producing an electric current. For biochemical energy, the fuel is biological material, such as enzymes, microbes, glucose, or marine sediment. These cells can be used in oceanographic systems and in human body implants, although the latter present challenges associated with biocompatibility and longevity. Glucose is present in the body, and therefore is biocompatible; however, some microbes are infectious. For glucose fuel cells implanted in dogs, power densities of approximately 2 $\mu W/cm^2$ have been reported [35]. Some researchers have endeavored to apply different fuel cells (*e.g.* from non-biochemical sources, such as methanol [36]) to power autonomous sensors, but these projects remain under development.

Since fuel cells are structurally similar to batteries, they can be modeled using electrical models similar to those used for batteries. The main difference is that fuel cells are conversion devices (*i.e.* they produce energy as long as fuel and oxidant are supplied), whereas batteries are storage devices (*i.e.* they contain limited energy) [37]. The voltage of a single cell in open circuit is typically around 1 V or less; thus, to achieve higher voltages, several cells are stacked in series.

3.6 Conclusions

This chapter covers the different types of ambient energy sources used for powering autonomous sensors: *radiant, mechanical, thermal, magnetic* and *biochemical*. Each energy source requires an appropriate energy transducer for converting the ambient or transferred energy into electrical energy.

Optical (or solar) energy, which is a type of radiant energy, offers the highest power density for outdoor applications. Indoors, energy mainly comes from artificial lights. Radiofrequency energy, which is also radiant, can be harvested from the environment or deliberately radiated. Available ambient RF energy is very small (*i.e.* in the nanowatt range), except at locations near RF transmitters. Industrial, scientific and medical (ISM) bands can be used to deliberately radiate RF energy.

Mechanical energy can be harvested via three types of interactions: *electrostatic, piezoelectric* and *electromagnetic*, each of which demands a different type of energy transducer. The past few years have witnessed intense research effort in this area, which has yielded several commercial harvesters, the majority of which are based on the electromagnetic principle.

The most widely used thermal transducer is based on thermoelectric materials, which offers better efficiencies than pyroelectric transducers. However, the efficiencies achieved at the temperature gradient between skin and air are still far from theoretical maximal values.

Magnetic energy can be found near ac current-carrying conductors or electric rotating machinery. Electromagnetic induction in a coil is the most widely harnessed principle for this type of energy and has been largely used in RFID systems and biomedical implants.

Biochemical energy can be extracted with fuel cells that employ biological materials (*e.g.* enzymes, microbes, glucose or marine sediment) as fuel. These cells can be used to power human implants, which may pose biocompatibility and longevity issues.

References

1. J.P. Thomas, M.A. Qidwai, J.C. Kellogg, Energy scavenging for small-scale unmanned systems. Journal of Power Sources **159**, 1494–1509 (2006)
2. S. Roundy, D. Steingart, L. Frechette, P. Wright, J. Rabaey, Power sources for wireless sensor networks, in *Proceedings of 1st European Workshop on Wireless Sensor Networks, EWSN* (2004), pp. 1–17
3. S.F.J. Flipsen, Power sources compared: The ultimate truth? Journal of Power Sources **162**, 927–934 (2006).
4. N. Mokhoff, IEDM feels the power. Online EETimes Europe (2006, October 27, 2006). Available on: http://www.eetimes.eu/193000825
5. J.F. Randall, *Designing Indoor Solar Products. Photovoltaic Technologies for AES*, 1st edn. (John Wiley & Sons, West Sussex, 2005)
6. T. Markvart, L. Castañer, *Practical Handbook of Photovoltaics: Fundamentals and Applications* (2003)
7. J.A. Hagerty, F.B. Helmbrecht, W.H. McCalpin, R. Zane, Z. B. Popovic, Recycling ambient microwave energy with broad-band rectenna arrays. IEEE Trans. on Microwave Theory and Techniques **52**, 1014–1024 (2004)
8. European Radiocommunications Office Recommendation, ERC RECOMMENDATION 70-03 (Tromsø1997 and subsequent amendments) RELATING TO THE USE OF SHORT RANGE DEVICES (SRD). Available on: http://www.erodocdb.dk/docs/doc98/official/pdf/REC7003E.PDF. Accessed March 27, 2008
9. J.-P. Curty, M. Declercq, C. Dehollain, N. Joehl, *Design and Optimization of Passive UHF RFID Systems*, 1st edn. (Springer Science+Business Media, New York, 2007)
10. J.P. Curty, N. Joehl, F. Krummenacher, C. Dehollain, M.J. Declercq, A model for u-power rectifier analysis and design. IEEE Trans. Circuits and Systems I: Regular Papers **52**, 2771–2779 (2005)
11. P.D. Mitcheson, E.M. Yeatman, G.K. Rao, A.S. Holmes, T.C. Green, Energy harvesting from human and machine motion for wireless electronic devices. Proceedings of the IEEE **96**, 1457–1486 (2008)
12. S. Roundy, J.M. Rabaey, P.K. Wright, *Energy Scavenging for Wireless Sensor Networks, with Special Focus on Vibrations*, 1st edn. (Kluwer Academic, Norwell, 2004)
13. S.P. Beeby, M.J. Tudor, N.M. White, Energy harvesting vibration sources for microsystems applications. Measurement Science and Technology **17**, R175 (2006)
14. N.S. Shenck, J.A. Paradiso, Energy scavenging with shoe-mounted piezoelectrics. IEEE Micro **21**, 30–42 (2001)
15. P.D. Mitcheson, E.M. Yeatman, G.K. Rao, A.S. Holmes, T.C. Green, Energy harvesting from human and machine motion for wireless electronic devices. Proceedings of the IEEE **96**, 1457–1486 (September 2008)
16. B.H. Stark, P.D. Mitcheson, M. Peng, T.C. Green, E.M. Yeatman, A.S. Holmes, Converter circuit design, semiconductor device selection and analysis of parasitics for micropower electrostatic generators. IEEE Trans. Power Electronics **21**, 27–37 (2006)
17. E.O. Torres, G.A. Rincon-Mora, Electrostatic energy harvester and Li-ion charger circuit for micro-scale applications, in *Proceedings of 49th IEEE International Midwest Symposium on Circuits and Systems, MWSCAS* (2006), pp. 65–69
18. E.O. Torres, G.A. Rincon-Mora, Energy-harvesting system-in-package microsystem. Journal of Energy Engineering **134**, 121–129 (2008)
19. P. Mitcheson, T. Green, E. Yeatman, Power processing circuits for electromagnetic, electrostatic and piezoelectric inertial energy scavengers. Microsystem Technologies **13**, 1629–1635 (2007)
20. S. Xu, K.D.T. Ngo, T. Nishida, G.B. Chung, A. Sharma, Low frequency pulsed resonant converter for energy harvesting. IEEE Trans. Power Electronics **22**, 63–68 (2007)
21. S. Roundy, P.K. Wright, J. Rabaey, A study of low level vibrations as a power source for wireless sensor nodes. Computer Communications **26**, 1131–1144 (2003)

22. C. Serre, A. Pérez-Rodríguez, N. Fondevilla, E. Martincic, S. Martínez, J. Morante, J. Montserrat, J. Esteve, Design and implementation of mechanical resonators for optimized inertial electromagnetic microgenerators. Microsystem Technologies **14**, 653–658 (2008)
23. A. Cuadras, M. Gasulla, V. Ferrari, Thermal energy harvesting through pyroelectricity. Sensors and Actuators A: Physical **158**, 132–139 (2010)
24. J. Xie, P.P. Mane, C.W. Green, K.M. Mossi, K.K. Leang, Energy harvesting via pyroelectric effect using PZT. Presented at Conference on Smart Materials, Adaptive Structures and Intelligent Systems, ASME Turf Valley Resort, Ellicott City, MD, USA, October 28–30
25. D. Guyomar, G. Sebald, E. Lefeuvre, A. Khodayari, Toward heat energy harvesting using pyroelectric material. Journal of Intelligent Material Systems and Structures **20**, 265–271 (2009)
26. R.J.M. Vullers, R. van Schaijk, I. Doms, C. Van Hoof, R. Mertens, Micropower energy harvesting. Solid-State Electronics **53**, 684–693 (2009)
27. G. Sebald, S. Pruvost, D. Guyomar, Energy harvesting based on Ericsson pyroelectric cycles in a relaxor ferroelectric ceramic. Smart Materials and Structures **17** (2008)
28. T. Taithongchai, E. Leelarasmee, Adaptive electromagnetic energy harvesting circuit for wireless sensor application, in *Proceedings of 6th International Conference on Electrical Engineering/Electronics, Computer, Telecommunications and Information Technology, ECTI-CON* (2009), pp. 278–281
29. A. Kurs, A. Karalis, R. Moffatt, J.D. Joannopoulos, P. Fisher, M. Soljacic, Wireless power transfer via strongly coupled magnetic resonances. Science **317**, 83–86 (2007)
30. K.v. Schuylenbergh, R. Puers, *Inductive Powering. Basic Theory and Application to Biomedical Systems*. Analog Circuits and Signal Processing Series (Springer, Berlin, 2009)
31. B. Lenaerts, Puers, Robert, *Omnidirectional Inductive Powering for Biomedical Implants*. Analog Circuits and Signal Processing Series (Springer, Berlin, 2009)
32. C.L. Zhang, J.S. Yang, W.Q. Chen, Harvesting magnetic energy using extensional vibration of laminated magnetoelectric plates. Applied Physics Letters **95**, 013511 (2009)
33. D.J. Dlugos, D. Small, D.A. Siefer, Wiegand effect energy generators, WO Patent 0017997, 2000
34. P. Woias, Y. Manoli, T. Nann, F.V. Stetten, Energy harvesting for autonomous microsystems, Mst News, 42–45 (2005)
35. S. Kerzenmacher, J. Ducrée, R. Zengerle, F. von Stetten, Energy harvesting by implantable abiotically catalyzed glucose fuel cells. Journal of Power Sources **182**, 1–17 (2008)
36. E.O. Torres, C. Min, H.P. Forghani-zadeh, V. Gupta, N. Keskar, L.A. Milner, I.P. Hsuan, G.A. Rincon-Mora, SiP integration of intelligent, adaptive, self-sustaining power management solutions for portable applications, in *Proceedings of the IEEE International Symposium on Circuits and Systems, ISCAS* (2006), pp. 5311–5314
37. EG&G Services Parson Inc., *Fuel Cell Handbook*, 5th edn. (U.S. Department of Energy, Morgantown, 2000)

Chapter 4
Primary Batteries and Storage Elements

As explained in Chap. 1, primary batteries and energy harvesters are the most common power sources for autonomous sensors. Storage units (secondary batteries and supercapacitors) are important for energy harvesters. Given their similarities, primary and secondary batteries are described jointly in this chapter. Supercapacitors are also explained, and their main differences compared to secondary batteries are highlighted. Simple circuit-models of batteries (both primary and secondary) and of supercapacitors are presented, the parameters of which were extracted through experimental characterization at different temperatures. Finally, the combined use of batteries and supercapacitors (*hybrid storage*), and its utility for autonomous sensors, is theoretically analyzed and experimental assessed.

4.1 Batteries

This section begins with a general overview of batteries, followed by an analysis of how an autonomous sensor's power consumption profile can affect battery voltage and performance. Finally, the most popular primary and secondary batteries are described.

4.1.1 General Characteristics

Batteries extract electrical power from a chemical reaction. They comprise one or more basic electrochemical units known as *cells*, which are connected in series or parallel to obtain the desired voltage and capacity. Each cell contains a negative electrode (an *anode*); a positive electrode (a *cathode*); and an ionic conductor (an *electrolyte*). The anode and cathode are physically isolated; the electrolyte provides the medium for charge transfer (via ions) inside the cell, between the anode and the cathode (Fig. 4.1(a)). In practical cells, the anode and the cathode are mechanically separated by a partition material that is permeable to the electrolyte [1].

M.T. Penella-López, M. Gasulla-Forner, *Powering Autonomous Sensors*,
DOI 10.1007/978-94-007-1573-8_4, © Springer Science+Business Media B.V. 2011

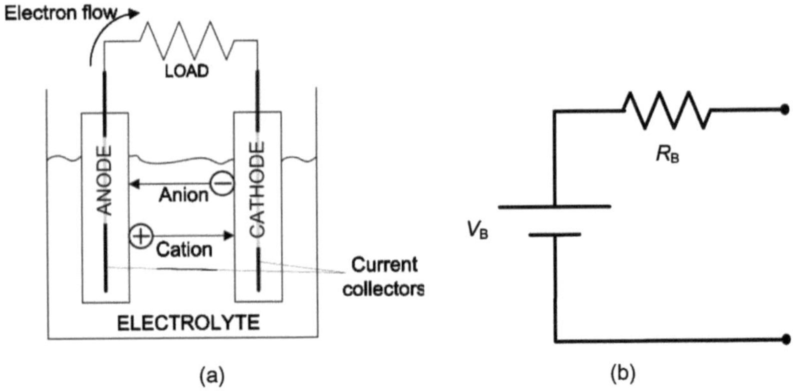

Fig. 4.1 (**a**) An electrochemical cell at discharge, and (**b**) a zero-order electrical model

A battery can be readily modeled with a zero-order electrical model similar to that shown in Fig. 4.1(b), whereby V_B models the open-circuit voltage (OCV), and R_B, the equivalent series resistance. The value of V_B decreases with decreasing internal stored charge (Q). In contrast, the value of R_B depends on the chemistry, battery design, and temperature. Section 4.2 introduces a more complex model for different types of batteries and explains its validation.

Most batteries are classified as either *primary* or *secondary*. Primary batteries cannot be electrically recharged readily; consequently, they are normally used once, and then discarded. They are commonly employed for simple home devices (*e.g.* remote controls, toys and clocks) and are the preferred option for long term, low-discharge current applications. Contrariwise, secondary batteries can be electrically recharged easily. Widespread use of portable devices (*e.g.* music players and mobile phones) has catalyzed the development of secondary batteries, for which numerous chemistries are now available.

All batteries, regardless of type or chemistry, share the following parameters and traits, which are described below [1, 2]: *capacity*; *voltage*; *energy density*; *power density*; *self-discharge*; *depth of discharge* (*DOD*); *state of charge* (*SOC*); and *temperature effects*.

4.1.1.1 Battery Capacity

A battery's capacity (C) is the total electrical charge that can be drained from a fully charged battery under specific discharge conditions (discharge rate, end voltage, and temperature). A battery's deliverable capacity is directly related to the amount of anode and cathode materials (*i.e.* active materials) that it contains. For secondary batteries, manufacturers usually report C in Ah. However, for primary batteries, they more often provide a discharge curve under a specified load (typically, a resistance); the discharge rate for primary batteries is usually expressed as the number of hours that the battery can power a certain load (*e.g.* a discharge rate of 10 h refers to the load that makes the battery last 10 h). Manufacturers use multiples or submultiples of *C* to express charge or discharge current. For example, a 100 mAh (1 C) battery discharged at 0.2 C (20 mA) will last 5 h. Obviously, this is not strictly correct, as *C* is a unit of charge, not of current. Nonetheless, this system remains widely used, and therefore, we have used it for this work. Additionally, commercial batteries have normalized sizes for a range of capacities [1].

4.1.1.2 Battery Voltage

The OCV of a battery depends on the standard potential of the active materials used and the amount of *Q* remaining. Lithium is widely used for anodes in primary and secondary batteries because it offers the highest standard potential among all battery chemistries.

A battery's discharge curve shows how the voltage delivered to a certain load changes versus remaining *Q* or versus time. This voltage also depends on the internal impedance of the battery and on the temperature. Some batteries present flatter discharge curves, as shown later in this chapter.

The nominal voltage is the voltage generally accepted as the typical value under discharge; however, it is not necessarily a representative point of the discharge curve.

4.1.1.3 Energy Density and Power Density

Energy density is the energy that can be extracted from a battery per unit volume (*volumetric*: Wh/l) or mass (*gravimetric*: Wh/kg). The practical energy density depends on the amount of active material inside the battery (and consequently, on *C*) and on the types of active materials (and consequently, on nominal voltage). Apart from active materials, a practical battery requires an electrolyte, current collectors, a separator, a container and terminals (see Fig. 4.1(a)). Thus, as greater volume or weight is dedicated to these components, rather than to active materials, the energy density will decrease. Some batteries are designed to improve energy density by providing more space for active materials (*e.g.* bobbin-type batteries).

Power density is the power that can be extracted from a battery per unit volume (*volumetric*: W/l) or mass (*gravimetric*: W/kg). It mainly depends on the internal impedance of the battery, which is tightly related to the current collector area and the electrolyte conductivity. Some batteries are specially designed to promote power density through reduced internal impedance; these include spiral-wound batteries, which maximize the current collector area. Batteries with high power density also have high self-discharge current. However, batteries designed to improve power density have less energy density for the same battery size (and *vice versa*). However, this tradeoff between energy and power density is more pronounced in small batteries. Small batteries (*e.g.* button-type) generally exhibit greater internal impedance than do larger ones with the same chemistry. Battery impedance varies with the extracted charge and increases with decreasing temperature. In secondary batteries, impedance usually increases with increasing number of cycles.

4.1.1.4 Self-discharge

Batteries self-discharge while stored on the shelf (*i.e.* unused) due to the occurrence of unwanted chemical reactions. *Self-discharge* (or *leakage*) increases with increasing temperature. Secondary batteries have much more self-discharge than primary batteries. For primary batteries self-discharge determines their lifetime (also known as *shelf life* for primary batteries). In contrast, the self-discharge of secondary batteries during storage has no permanent effect on their lifetime, as they can simply be recharged before use. Nevertheless, self-discharge can be as important as the power consumption of the autonomous sensor.

4.1.1.5 Depth of Discharge and State of Charge

A battery's *depth of discharge* (*DOD*) indicates how much of its charge has been extracted. Conversely, its *state of charge* (*SOC*) indicates how much charge remains. However, to obtain meaningful data on these parameters, the battery must be left unconnected for a sufficiently long period of time.

4.1.1.6 Temperature Effects

The rate of chemical reactions increases with increasing temperature: as a general rule, reaction rate doubles at every increase of 10 °C [2]. This principle has beneficial consequences (*e.g.* batteries show better performance because their resistance decreases) as well as detrimental ones (*e.g.* reduction of battery lifetime due to higher leakage). The lower limit for a battery's operating temperature range is partially determined by the electrolyte's freezing point.

4.1.1.7 Differences Between Primary and Secondary Batteries

The main difference between primary batteries and secondary batteries is that the latter can be electrically recharged, whereas the former cannot. Furthermore, the term *lifetime* has a different meaning for each type: as mentioned above, the lifetime (or *shelf life*) of primary batteries is limited by their self-discharge, and is usually defined as the time required for the battery to reach 90% of its original capacity. In contrast, the lifetime of secondary batteries is defined as the duration of satisfactory performance measured in years (*float* or *calendar life*) or in the number of charge/discharge cycles (*cycle life*). The end-of-life criterion for secondary batteries is typically defined as the time required for the battery to reach 80% of its original capacity. The number of charge/discharge cycles typically increases with decreasing DOD. Float life and cycle life are both affected by numerous conditions, including the temperature during charge and discharge; DOD; charge current and discharge current; the charge control method; exposure to overcharge and/or to overdischarge; and the storage conditions and duration.

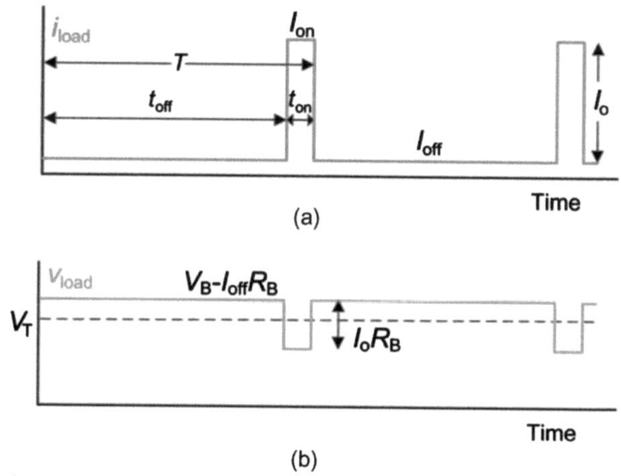

Fig. 4.2 (**a**) Current consumption in the sensor node, and (**b**) the resulting battery-voltage profile

Secondary batteries generally have lower energy density than primary batteries, as they require designs that facilitate recharging and maximize cycle life. For example, secondary batteries demand a safety vent for gas release in the event of high internal pressure build-up. They also exhibit greater self-discharge, and consequently, they are not recommended for use as the chief power source for autonomous sensors.

4.1.2 Batteries and Autonomous Sensors

As explained in Chap. 2, autonomous sensors are basically low-power pulsed loads. The load model presented in Chap. 2 can be simplified to consider only the current consumption during sleep (I_{off}) and an average current consumption while active (I_{on}). If this pulsed load is connected to a battery, the load voltage (v_{load}) will drop during the active time, because of the internal impedance of the battery (R_B in Fig. 4.1(a)). This situation is illustrated in Fig. 4.2, whereby T, t_{on}, and t_{off} are the period, the active time interval, and the sleep time interval, respectively, of the pulsed current (i_{load}), and $I_o = I_{on} - I_{off}$. The average current is $I_{c,av} = DI_o + I_{off}$, whereby D is the duty cycle of the autonomous sensor ($D = t_{on}/T$), which can be adjusted—whenever allowed by the application—to obtain the desired average current. Albeit I_{off} is several orders of magnitude lower than I_{on}, its contribution to $I_{c,av}$ is dominant in applications in which $D < I_{off}/I_{on}$. R_B and the associated voltage drop have some undesired effects on the autonomous sensor (mainly in the lifetime) that are more pronounced at higher values of R_B. For example, autonomous sensors will stop working when v_{load} falls below the minimum allowed supply voltage (V_T in Fig. 4.2(b)). Additionally, some power (and therefore, energy) is dissipated on R_B during t_{on}, thereby reducing the efficiency when energy is extracted from the battery. Thus, batteries with low R_B are preferred as the main power source for autonomous sensors.

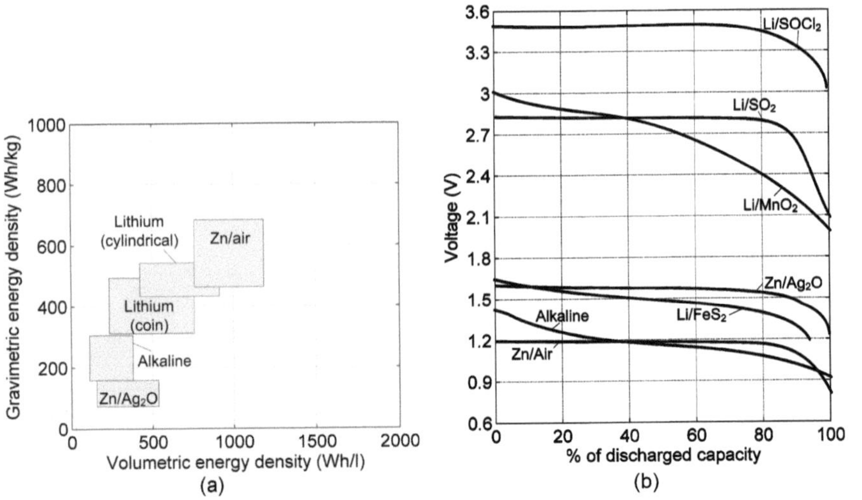

Fig. 4.3 (**a**) Gravimetric energy density plotted against volumetric energy density for common primary batteries, and (**b**) voltage discharge profiles at discharge rates of 30 h to 100 h, for the most common types of primary batteries. Data from [1–6]

4.1.3 Primary Batteries

Primary batteries fall into four categories, based on their chemical reagents [1, 3]:

- Alkaline (Zn/alkaline-MnO_2): these are the most common primary batteries, and are used in everyday applications such as TV remote controls, calculators, and toys.
- Lithium based (Li/XX): used in toys or medical devices that demand high reliability, long life and low self-discharge.
- Zinc-air (Zn/air): mostly used in hearing aids.
- Silver-oxide (Zn/Ag_2O): mainly used in watches.

Alkaline, zinc-air and silver-oxide batteries have zinc-based anodes, which offer good electrochemical performance, reasonably good shelf life, and low cost. In contrast, lithium batteries provide the highest standard potential; thus, batteries using a lithium anode offer higher cell voltages.

Figure 4.3(a) shows the gravimetric energy density versus the volumetric energy density of the four main types of primary batteries. Although lithium material offers the highest energy density, zinc-air batteries feature higher gravimetric and volumetric energy densities, owing to their air based cathode, which is not contained inside the cell, thereby providing more space for the anode. Figure 4.3(b) shows the voltage discharge profiles for the most common types of batteries. Zinc-air, silver-oxide, and some lithium batteries exhibit a flat discharge profile.

The principal characteristics of common primary batteries are summarized in Table 4.1. Internal impedance in these batteries is usually measured at 1 kHz. Although zinc-air batteries offer high energy density, their high internal impedance

Table 4.1 Principal characteristics of the four main types of primary batteries [1, 3]

Battery type	Alkaline	Lithium based	Zinc-air	Silver-oxide
Internal impedance (mΩ)	Low (tens)	Moderate (hundreds)	High (thousands)	High (thousands)
Discharge profile (Fig. 4.3(b))	Sloped	Flat/Sloped	Flat	Flat
Shelf life or lifetime (years)	4 to 5	10 to 20	3 to 4	4 to 5
Operating temperature (°C)	−20 to 55	−20 to 55[a]	0 to 50	0 to 55

[a]Most-restrictive range. $LiSOCl_2$ can work from −60 °C to 85 °C.

makes them generally unsuitable for autonomous sensors. Furthermore, they are only constructed in button configuration, with capacities less than 1 Ah. Additionally, since zinc-air batteries are open to air (which is their cathode material), they are more susceptible to environmental changes (*e.g.* in temperature or humidity). Silver-oxide batteries are quite expensive, only available in small capacities and also suffer from rather high internal impedance. Lithium based batteries afford the best tradeoff between energy density and internal impedance, but are typically quite expensive. Finally, alkaline batteries are moderately priced, but feature relatively low energy density and a sloped discharge profile. More information on these batteries types can be found in [1] to [4], among other references.

Lithium-iodine batteries are a special type of lithium based batteries intended for long-term microampere discharge, owing to their relatively high resistance. Indeed, their lifetime can extend up to 25 years. They feature small battery capacities of up to 0.5 Ah and can operate from 0 °C to 100 °C. They are typically used in pacemakers. Lithium thionyl chloride batteries are intended for applications with very low continuous current and moderate pulse current. They can operate up to 20 years and in a wide temperature range (−55 °C to 155 °C) [5].

4.1.4 Secondary Batteries

The most widely used secondary batteries for portable and space-constrained devices are [1, 2, 6]:

- Lithium based (Li-ion or Li polymer)
- Nickel metal hydride (NiMH)

Figure 4.4(a) shows a plot of volumetric energy density versus gravimetric energy density for typical secondary batteries. Metallic-lithium based batteries offer better energy densities but present more safety risks. Batteries using lithium alloys in the negative electrode (*e.g.* button $LiAl/MnO_2$ batteries) imply less risks, but at the expense of lower energy density and poor cycle life at deep DOD. These batteries are sold in coin configuration and are intended for low rate discharges.

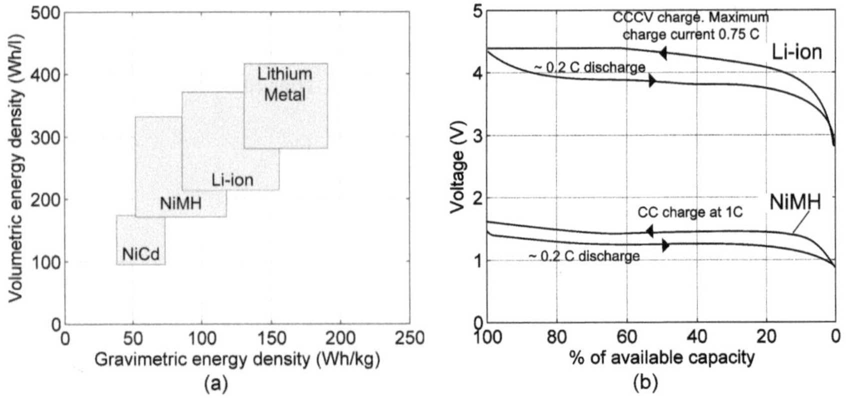

Fig. 4.4 (**a**) Volumetric energy density plotted against gravimetric energy density for common secondary battery types (data from [1–4, 6]); (**b**) charge (at 1 C and 0.75 C) and discharge (at 0.2 C) curves for Li-ion and NiMH batteries

Lithium-ion batteries use lithium intercalation compounds as positive and negative materials: their name comes from the exchange of lithium ions that occurs between the positive and negative electrodes as the battery is cycled. Over the past decade, they have become the most popular lithium based batteries, as they are relatively safer and offer reasonably high energy densities. A recently marketed Li-ion cell, based on lithium iron phosphate ($LiFePO_4$) as cathode, promises higher cycle life (the manufacturers claim up to 2000 cycles [7, 8]), fewer safety issues, and more tolerance to overcharge; however, its energy densities and cell voltages are relatively small (nominal voltage = 3.2 V). In terms of other parameters, such as internal impedance and operating temperatures, it resembles standard Li-ion batteries. Commercial $LiFePO_4$ cells have been designed for the electric car industry and have relatively high capacities (in most cases, > 5 Ah).

Figure 4.4(b) shows the charge and discharge curves for Li-ion and NiMH batteries. As observed in the plots, during charging, the battery voltage is higher than during discharge, which can be easily explained by the effect of internal impedance. Lithium-ion batteries present higher voltage ranges than do NiMH batteries.

The principal characteristics of the three most common types of secondary batteries are listed in Table 4.2. Lithium-polymer batteries are Li-ion batteries that employ polymer-based electrolytes, which enable slimmer, and even flexible, designs. They can be constructed in myriad shapes to fit the available space. The recommended charge method depends on the chemistry and can be a determinant in designing the energy conditioning block of autonomous sensors. The voltage discharge profiles for these batteries typically show a higher slope than those of primary batteries.

Secondary batteries are very sensitive to overcharge and overdischarge, and therefore, typically require protection circuits. If the battery voltage drops below the end-of-discharge voltage, it can be recovered by cycling the battery at low rates (0.1 C) several times. When charging Li-ion and Li polymer batteries, the voltage is

Table 4.2 Principal characteristics of the three main types of secondary batteries [1–3]

Battery type	Li-ion	Li polymer	NiMH
Internal impedance Z (mΩ)	30 to thousands	40 to thousands	50 to thousands
Discharge profile (Fig. 4.4(b))	Sloped	Sloped	Sloped
Recommended charge method	Constant-current constant-voltage (CCCV)	CCCV	Constant current (CC) with charge termination
Over-voltage tolerance	Explosion risk	Damage risk	Temperature increase
Temperature range (°C)			
Charge	0 to 40	0 to 45	0 to 45[a]
Discharge	−20 to 60	−20 to 60	−20 to 60[a]
Self-discharge (% per month)	2	2.5	15 to 25
Cycle life (cycles)	500 to 1000	300 to 500	300 to 500
Float or calendar life (years)	> 5	> 5	4 to 6

[a]These are the permissible temperature ranges; however, the cell should operate as closely to RT as possible.

limited to 4.1 V or 4.2 V for safety reasons. Due to their inherent safety risks, Li-ion batteries are typically sold with a built-in protection circuit that disconnects the cell in the event that charging current or cell temperature is too high. Their operating temperature range depends on whether they are being charged or discharged. Their self-discharge is rated at room temperature (RT; = 25 °C ± 1 °C); it increases with increasing temperature. Despite their higher initial cost relative to primary batteries, secondary batteries offer a cheaper per-cycle cost. Lithium based batteries are typically more expensive than NiMH batteries. Despite not having memory effects, NiMH batteries should be completely cycled every few months.

A battery pack must always comprise batteries of the same capacity, SOC and cycle life in order to avoid polarity reversal in any one of the cells. Furthermore, active or passive circuits can be used to maintain the voltage of the individual cells within the same range, although this wastes some power. This method is called *cell balancing*.

Thin-film batteries are solid-state lithium batteries specially designed for integration onto silicon wafers, as they are very thin (micrometers). Both Li-ion and metallic lithium thin-film batteries have been developed. This technology could theoretically allow every integrated circuit to have an independent built-in power source. They can sustain high temperatures (250 °C) for short periods of time and contain no liquid elements; therefore, they can support solder reflow [1]. Cycle lives of several thousands of cycles at RT and hundreds of cycles at 150 °C have been demonstrated [9]. These batteries exhibit energy densities of 100 μWh/cm^2 to 1000 μWh/cm^2, depending on the cathode [10], and power densities (under continuous discharge) of up to 10 mW/cm^2 [9]. Therefore, their internal impedance is quite high compared to traditional Li-ion cells, which they otherwise resemble closely. There are companies that work on and commercialize these types of batteries, *e.g.* Excellatron, Cymbet, and Infinite Power Solutions.

Charging of secondary batteries from the electrical grid uses some circuitry to protect the battery. These circuits deliver the appropriate voltage and current to the battery depending on its DOD. When harvesting energy from the environment, the charging current is usually quite low, thereby minimizing the need for complex charging and protection circuits. Simple over or under voltage protection circuits (*e.g.* a comparator plus a disconnection switch) can be considered. More information on battery protection circuits can be found in [11].

4.2 Battery Characterization

Internal impedance in batteries depends on factors including the battery chemistry, the cell size and construction, the number of connected cells, the wiring, and the contact type [6]. Manufacturers normally provide graphs of a battery's voltage plotted against its remaining capacity for different constant loads (as in Figs. 4.3(b) and 4.4(b)), as well as a value for ac impedance at a specific frequency and DOD. However, the data found in these plots are not sufficient for predicting a battery's voltage behavior under pulsed loads, such as that of autonomous sensors (Fig. 4.2(a)). Thus, prediction of the transient response of the supply voltage of low-power autonomous sensors requires a suitable electrical model of the batteries. This model can be useful for predicting the runtime of autonomous sensors, because the presence of excessive voltage drops (see Sect. 4.1.2), and for measuring the battery OCV to easily determine the battery DOD. Knowing the transient response of the battery enables determination of when the OCV has been reached.

Among the different models proposed for batteries, electrical models seem to be the most convenient. Min and Rincon-Mora [12] have proposed a model that incorporates OCV to simulate transient response using two resistor-capacitor (RC) networks. Herein, we describe a simplified version of this model, which uses only a single RC network. We determined the value of the model parameters from experimental data by draining a pulse of current from the batteries, thus emulating pulsed loads. We tested different types of low capacity (< 1.5 Ah) primary and secondary batteries, and then compared experimental transient responses against simulation data using the parameter values determined for the batteries.

4.2.1 Proposed Approach

Figure 4.5 shows the battery model proposed to predict the transient response of the battery voltage under pulsed current loads. This model is commonly used for

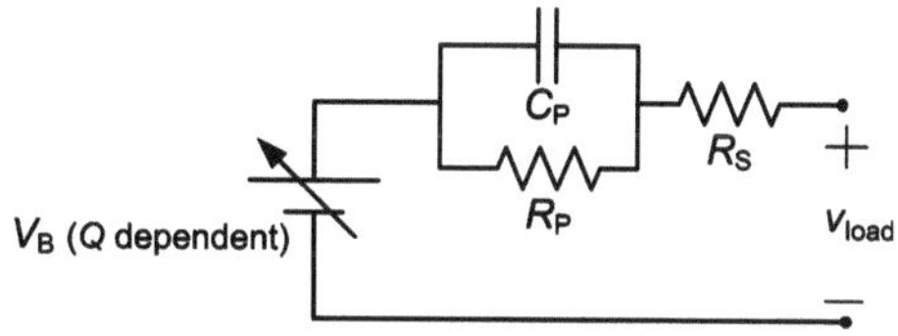

Fig. 4.5 Proposed battery model

electrochemical impedance measurements and is a simplified version of a Randles cell [13]. The voltage source (V_B) models the OCV, which depends on Q; the series resistance (R_S) stems from the ohmic contacts of the current collectors and electrodes; the parallel resistor (R_P) models the ionic resistance of the electrolyte; and the capacitor (C_P) models the double-layer effect that appears at the interface between the electrodes and the electrolyte [1]. The voltage at the battery terminals is equal to the voltage available to the load (v_{load}).

To estimate the parameters of the model (R_S, R_P and C_P), we theoretically analyzed how v_{load} varies while a single current pulse (i_{load}) is drained from the battery. The battery impedance in the Laplace domain is given by:

$$Z_B(s) = \frac{R_S \tau s + R_S + R_P}{\tau s + 1}, \tag{4.1}$$

whereby

$$\tau = R_P C_P. \tag{4.2}$$

The current pulse drained from the battery can be expressed in the temporal and Laplace domains as shown below in (4.3) and (4.4), respectively:

$$i_{load}(t) = I_o\big(u(t) - u(t - t_{on})\big), \tag{4.3}$$

$$I_{load}(s) = I_o\left(\frac{1}{s} - \frac{e^{-t_{on}s}}{s}\right), \tag{4.4}$$

whereby $u(t)$ is the step function, and the pulse starts at $t = 0$ and ends at $t = t_{on}$. This i_{load} is like a single pulse in Fig. 4.2(a) with $I_{off} = 0$; consequently, $I_o = I_{on}$. An $I_{off} \neq 0$ could easily be considered by adding a constant current to i_{load}. v_{load} changes according to:

$$\begin{aligned} v_{load}(t) &= V_B - I_o(R_S + R_P)\left\{\left[1 - \frac{R_P}{R_S + R_P} e^{-\frac{t}{\tau}}\right] \times u(t) \right. \\ &\quad \left. - \left[1 - \frac{R_P}{R_S + R_P} e^{-\frac{(t-t_{on})}{\tau}}\right] \times u(t - t_{on})\right\} \\ &= V_B - I_o\big\{R_S + R_P\big(1 - e^{-\frac{t}{\tau}}\big) \times u(t) \\ &\quad - \big[R_S + R_P\big(1 - e^{-\frac{(t-t_{on})}{\tau}}\big) \times u(t - t_{on})\big]\big\}. \end{aligned} \tag{4.5}$$

Figure 4.6(a) shows a temporal representation of (4.5). The instantaneous voltage changes at $t = 0$ and $t = t_{on}$ are due to R_S, whereas the exponential behavior during the remaining time is due to R_P and C_P. Values for R_S, R_P and C_P can be estimated from (4.5) with at least three different points. We chose the points marked in

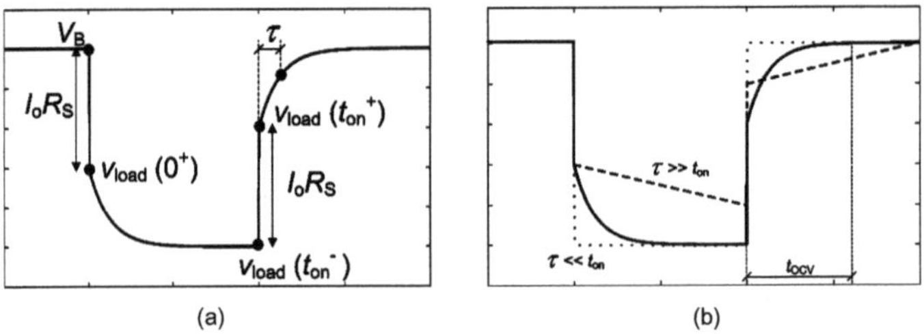

Fig. 4.6 Temporal representation of (4.5): (**a**) points chosen for estimating R_S, R_P and C_P; and (**b**) voltage behavior as a function of τ

Fig. 4.6(a). We first derived R_S from the initial voltage drop at $t = 0$; subsequently determined τ from the graph; and finally, obtained R_P and C_P. The analytical expressions corresponding R_S, to R_P and C_P are shown in (4.6) to (4.8), below.

$$R_S = \frac{V_B - v_{load}(0^+)}{I_o}, \tag{4.6}$$

$$R_P = \frac{\frac{V_B - v_{load}(t_{on}^-)}{I_o} - R_S}{(1 - e^{-\frac{(t_{on})}{\tau}})} = \frac{v_{load}(0^+) - v_{load}(t_{on}^-)}{I_o(1 - e^{-\frac{(t_{on})}{\tau}})}, \tag{4.7}$$

$$C_P = \frac{\tau}{R_P}. \tag{4.8}$$

In Fig. 4.6(b), t_{OCV} accounts for the time that must be waited following deactivation of the pulse current before measuring the OCV (V_B) with a bounded error. The lower the permissible error, the longer the t_{OCV}. The dotted and dashed lines represent the cases $\tau \ll t_{on}$ and $\tau \gg t_{on}$, respectively. A higher τ implies a longer t_{OCV} but lower voltage drops.

When I_{off} is not negligible, the measured value at $t = 0^-(V_B)$ equals the OCV minus $I_{off}(R_S + R_P)$; however, R_S, R_P and C_P are estimated using the same equations as explained above. This assumption holds true whenever I_{off} is present long before $t = 0$. Under these circumstances, the OCV—and consequently, the DOD—can only be obtained if I_{off} is known. The points used to compute R_S, R_P and C_P can be chosen such that the algorithm could be automated or even implemented in a low power microcontroller by using simple equations [14]. Thus, if the current behavior of the autonomous sensor is well characterized (Fig. 4.2(a)), then R_S, R_P and C_P can be extracted from the behavior of v_{load} at the activation time of the autonomous sensor. In this case, battery information can be extracted without disconnecting the autonomous sensor. However, proper measurement demands that T and t_{on} be tailored to the expected τ.

4.2.2 Materials and Methods

We evaluated nine different batteries (four primary and five secondary) representing the most common technologies at RT. Table 4.3 lists the models (the first four

Table 4.3 Tested batteries and their main characteristics

Battery (manufacturer)	Nominal capacity (mAh)	Nominal voltage (V)	Impedance @ 1000 Hz (mΩ) (manufacturer data)
V357 Silver-oxide (Varta)	165	1.55	Not specified
675 Zinc-air (Duracell)	600	1.4	3000
E92 Alkaline (Energizer)	1250	1.5	150–300[b]
CR2 Lithium (Panasonic)	1400	3	Not specified
ML 2016 Lithium (Maxell)	25	3	Not specified
GP17AAAH2X[a] NiMH (GP)	170	2.4	< 400
NP FE1 Lithium-ion (Energizer)	450	3.6	Not specified
Lithium-polymer (Varta)	550	3.7	130
GP130AAM NiMH (GP)	1300	1.2	17–38

[a]Battery pack consisting of two cells connected in series.

[b]Measured at dc with the dual pulse method: a 100 ms current pulse is drained from the battery, and the resistance is then estimated from the voltage drop at the end of the pulse.

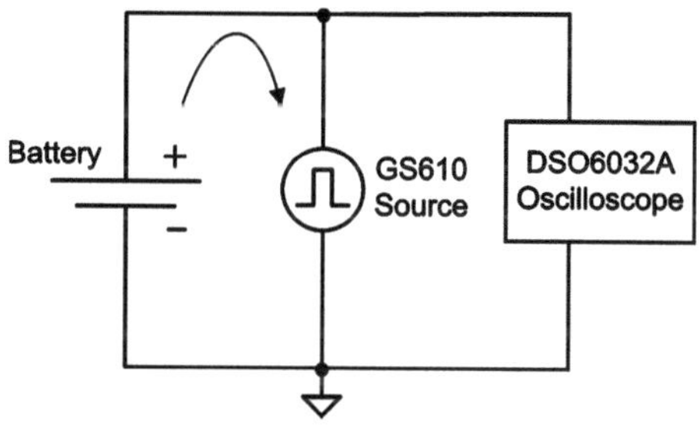

Fig. 4.7 Measurement setup

are primary, and the last five, secondary) with their principal characteristics. The nominal capacities ranged from 25 mAh to 1400 mAh. A manufacturer-provided impedance value (generally measured at 1 KHz) was only available for five of these batteries.

A pulsed current ($T = 10$ s and $t_{on} = 0.1$ s) was generated using a GS610 source (Yokogawa). A larger t_{on} (3 s) was used for the silver-oxide battery because of its larger measured τ. Amplitudes (I_o) from 30 mA to 1 A were employed (depending on the internal battery impedance) to produce a measurable voltage drop, which was acquired with a DSO6032A oscilloscope (Agilent). The values of R_S, R_P and C_P were then calculated using (4.6) to (4.8). Figure 4.7 shows the measurement setup.

Some of the batteries were tested at 0 °C and −20 °C by using a FCH2730 climatic chamber (CCI). The results for two of these batteries are shown in Figs. 4.9 and 4.10. When the temperature was set to 0 °C, the internal temperature of the climatic chamber, measured with the thermo-hygrometer DO 4096 (Delta Ohm), oscillated between −1.9 °C and 3.3 °C; in contrast, when the temperature was set to −20 °C, the internal temperature oscillated between −15 °C and −19.7 °C. The relative humidity fluctuated between 22% and 56%.

Table 4.4 Measured values for R_S, R_P and C_P at RT and full charge for the primary batteries tested

Battery	Nominal capacity (mAh)	Current applied (mA)	R_S (Ω)	R_P (Ω)	C_P (mF)
V357 Silver-oxide (Varta)	165	50	3.26	1.403	361
675 Zinc-air (Duracell)	600	300	0.863	3.38	79
E92 Alkaline (Energizer)	1250	500	0.134	0.026	5450
CR123A Lithium (Panasonic)	1400	500	0.35	0.096	320

Table 4.5 Measured values for R_S, R_P and C_P at RT and full charge for the secondary batteries tested

Battery	Nominal capacity (mAh)	Current applied (mA)	R_S (Ω)	R_P (Ω)	C_P (mF)
ML 2016 Lithium/MnO_2 (Maxell)	25	30	11.24	8.31	1.7
GP17AAAH2X NiMH (GP)	170	200	0.481	0.524	135
NP FE1 Lithium ion (Energizer)	450	1000	0.084	0.171	76
Lithium polymer (Varta)	550	1000	0.119	0.125	217
GP130AAM NiMH (GP)	1300	500	0.104	0.076	184

4.2.3 Experimental Results

Table 4.4 shows the R_S, R_P and C_P (Fig. 4.5) obtained for the four tested primary batteries at RT and full charge. As observed in the table, zinc-air and silver-oxide batteries, which have lower capacities, showed higher internal impedance. For the zinc-air and alkaline batteries, the sum of R_S and R_P is close to the impedance value provided by their respective manufacturer (Table 4.3). However, no manufacturer's data were available for the other two primary batteries tested.

Table 4.5 shows the parameter values for modeling the five tested secondary batteries. As observed in the table, the batteries with the lowest capacities gave the highest R_S and R_P (see Sect. 4.1.1.3, [15]).

The ML2016 battery was also tested at RT for different OCVs (Fig. 4.8). As observed in the figure, resistance increased with decreasing OCV, and thus, with increasing DOD, which is standard behavior [16]; C_P and τ also increased with increasing DOD.

For the 675 zinc-air primary battery (Fig. 4.9) and the NP FE1 Li-ion secondary battery (Fig. 4.10), resistance and capacitance were also measured at full charge and at 0 °C and −20 °C. As expected, resistance increased with decreasing temperature [16].

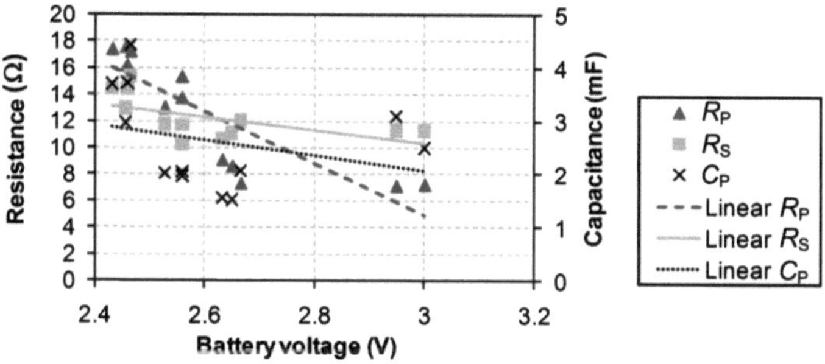

Fig. 4.8 Calculated parameters versus OCV for the ML2016 lithium battery at RT

Fig. 4.9 Resistance and capacitance at different temperatures and full charge for the 675 zinc-air battery

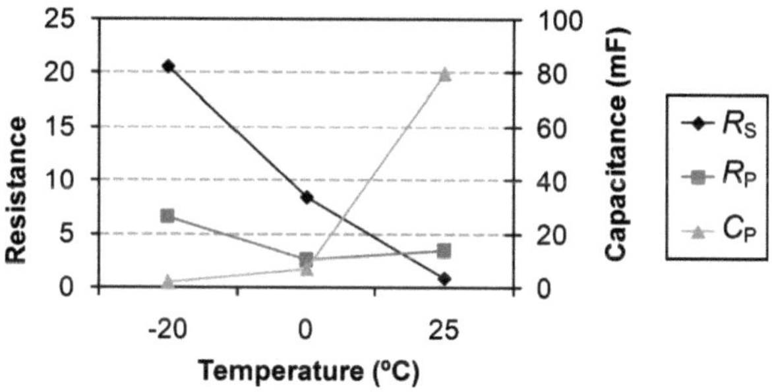

Fig. 4.10 Resistance and capacitance at different temperatures and full charge for the NP FE1 Li-ion battery

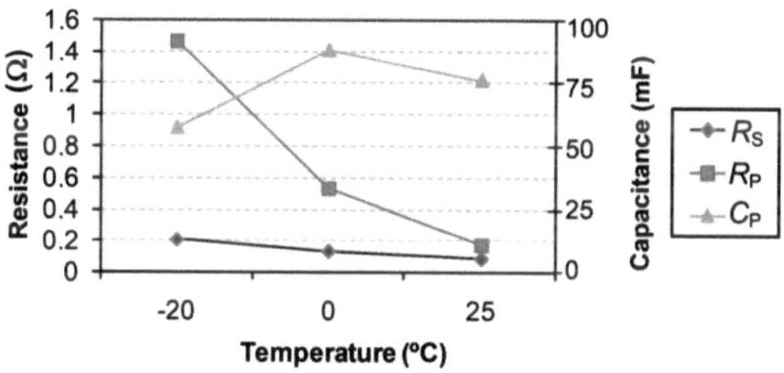

4.2.4 Model Validation

To validate the proposed model shown in Fig. 4.5, we compared the transient voltage response of six of the tested batteries against simulated data (Matlab). For the simulations, we used the R_S, R_P and C_P values from Tables 4.4 and 4.5, and a current pulse with the same amplitude and width as that used in the experimental tests. Table 4.6 shows the root mean square error (ε_{rms}) computed using (4.9) for each battery. This error was less than 3% for all the batteries except the zinc-air one.

Table 4.6 Model validation results

Battery	ε_{rms} (%)
ML 2016 Lithium (Maxell)	2.99
Lithium-polymer (Varta)	2.01
E92 Alkaline (Energizer)	1.41
CR2 Lithium (Panasonic)	1.83
675 Zinc-air (Duracell)	11.7
V357 Silver-oxide (Varta)	2.43

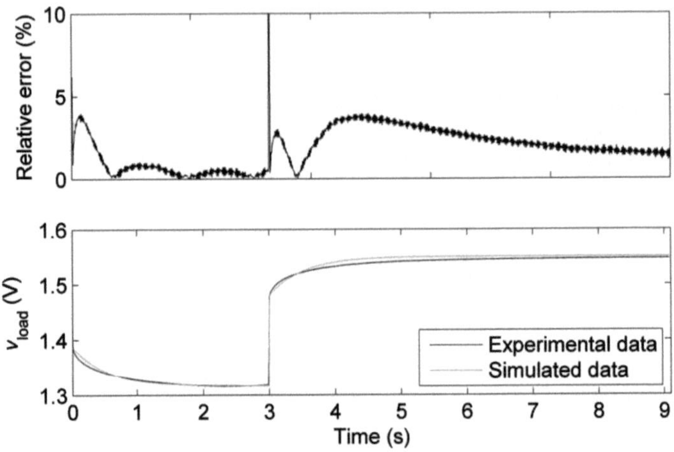

Fig. 4.11 Relative error, and acquired and simulated data, at $t_{on} = 3$ s and $I_o = 50$ mA for the silver-oxide battery tested

This may be because the amplitude of the current pulse used in the experimental characterization was too high, leading to more erroneous characterization.

$$\varepsilon_{rms}(\%) = 100\sqrt{\frac{1}{N}\sum_{k=1}^{N}\left(\frac{(v_{simulated} - v_{experimental})^2}{[V_B - v_{load}(t_{on}^-)]^2}\right)}, \tag{4.9}$$

Figure 4.11 shows the relative error for the silver-oxide battery and the conditions at which the model was fitted. As observed in the figure, a relatively high error occurs at $t = 3$ s, right when the pulse current is switched off. For the remaining data, the acquired and simulated voltage behaviors are quite similar, and the relative error is less than 4%.

4.3 Supercapacitors

Supercapacitors, also called *ultracapacitors* or *double-layer capacitors*, feature activated carbon electrodes that have very large surface areas and that may be sepa-

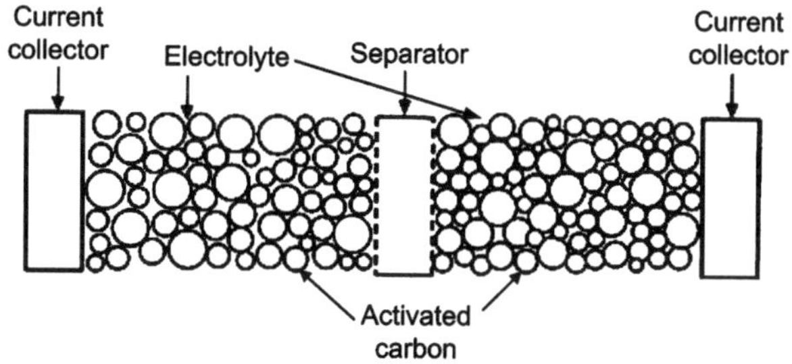

Fig. 4.12 Structure of a supercapacitor. Adapted from [18]

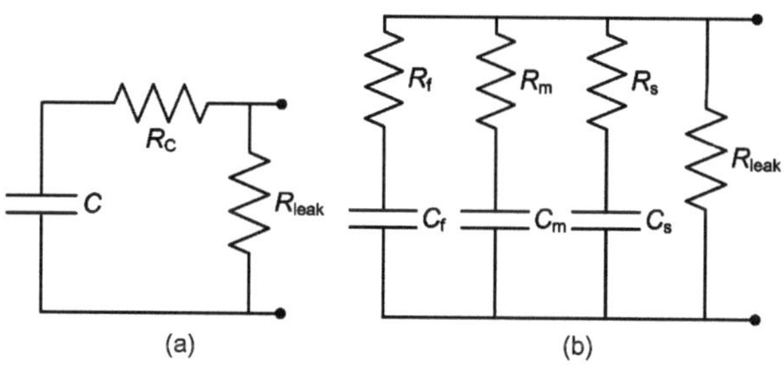

Fig. 4.13 (**a**) A simple supercapacitor circuit model; and (**b**) a more complex supercapacitor circuit model [20]

rated by distances as short as in the molecular range. These characteristics enable higher capacities per unit area than that of regular capacitors. Moreover, in contrast to regular capacitors, supercapacitors do not have a dielectric between their two electrodes and charge/discharge occurs in the ion absorption layer formed on the electrode surfaces (see Fig. 4.12). Although their two electrodes are equal, in order to preserve lifetime, the polarity must not be reversed [17]. The electrodes are impregnated with an electrolyte that enables charge transfer via ions. There are two types of electrolytes: *organic* and *aqueous*. Organic electrolytes provide higher voltage per cell and higher energy densities, whereas aqueous electrolytes lead to smaller cell-to-cell voltage differences.

Supercapacitors are often used for auxiliary applications: for example, to provide backup power for computer or cell phone memory, or to generate brief bursts of energy in battery-powered consumer products (*e.g.* cameras) [19].

Figure 4.13(a) shows a simple electrical model for a supercapacitor, which is also used for regular capacitors, whereby C is the nominal value of the capacitor; R_C models the equivalent series resistance (ESR); and R_{leak} models the supercapacitor leakage current. There are more complex supercapacitor models, such as one which includes several RC branches in parallel [20]. Figure 4.13(b) shows an example with three RC branches and a parallel R_{leak} that models the supercapacitor leakage. The main branch ($R_f C_f$) is equivalent to CR_C in Fig. 4.13(a). The other RC branches model longer time constants (from seconds or minutes to days). Consequently, fully charging or discharging a supercapacitor may take several hours or days. Obviously, the *slower* RC branches cannot store or deliver energy as quickly as the main branch. However, C_f is much larger than the model's remaining capacitors, and consequently, so is its corresponding available energy. The leakage current is the current that passes through R_{leak} once the capacitor has been fully charged

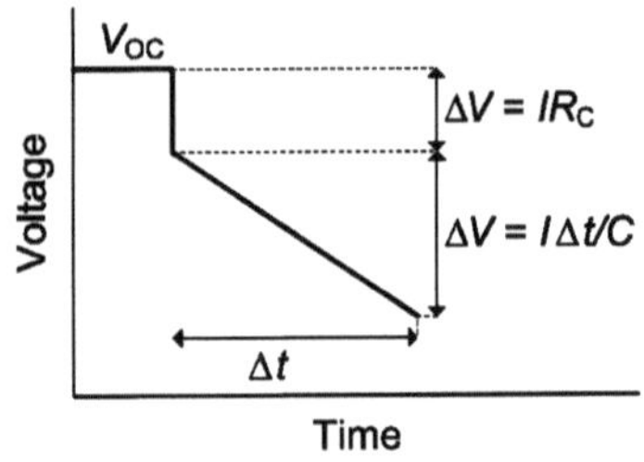

Fig. 4.14 Supercapacitor discharge under a step current load [23]

or discharged (*i.e.* the capacitors of all the branches) to some fixed voltage. Other complex models of supercapacitors can be found in [21] and [22].

Figure 4.14 shows the voltage behavior of a supercapacitor under a step current load simulated using the simplified model in Fig. 4.13(a). First, R_C causes an initial voltage drop when the load is applied, and then the voltage of the supercapacitor decreases linearly (as happens with regular capacitors). Supercapacitors designed for backup applications have a relatively high R_C (large voltage drop for a step current load), but also have a high R_{leak} (low leakage). Conversely, supercapacitors with very low R_C (tens to hundreds of milliohms) have a relative low R_{leak} (high leakage).

For supercapacitors, size and leakage current increase with nominal capacity. Leakage current also increases with increasing temperature (similarly to batteries) and with increasing voltage. Leakage current must be considered for applications with low average current consumption. R_C also increases at low temperatures, although not considerably [21].

The maximum voltage that can be applied to a supercapacitor is determined by the breakdown of the solvents in the electrolyte. The maximum voltage and the rated charge/discharge current must not be exceeded; apart from these considerations, no other special care must be taken for charging/discharging. When using stacks of supercapacitors connected in series, the potential differences between them can cause the voltage in one supercapacitor to exceed the maximum allowed voltage. Thus, to keep the voltage of all the supercapacitors within the safe range, voltage balancing must be considered [23]. This is usually done with a balancing resistance placed in parallel with the supercapacitor; however, this balancing resistance ultimately increases supercapacitor leakage. The company AVX claims that supercapacitors with aqueous electrolytes do not require balancing, and that these can easily operate at higher voltage, with no added leakage, by stacking several cells in series [24]. They manufacture supercapacitors with nominal voltages of up to 16 V.

Table 4.7 summarizes the most important characteristics of supercapacitors. Compared to secondary batteries (Table 4.2), supercapacitors feature energy densities of one to two orders of magnitude lower. However, their power densities can be much higher (and conversely, their internal impedances can be much lower). Supercapacitors have an unlimited shelf life when stored in a discharged state and low ageing when floating at nominal voltage. Additionally, their DOD can be easily obtained by measuring their OCV. However, the voltage decreases linearly with the DOD, which poses more challenges to the preceding energy conditioning stage

Table 4.7 Main characteristics of supercapacitors [17, 18, 23, 25, 26]

Capacity	Units of μAh—hundreds of mAh (units of mF—hundreds of F)
Energy density (mWh/cm^3)	1–10
Series resistance	Units of mΩ to tens of Ω
Leakage	Units or tens of μA
Maximum voltage (V/cell)	2.3–2.75 (organic electrolyte) 1 (aqueous electrolyte)
Charge characteristics	Maximum voltage must not be exceeded.
Discharge characteristics	Linear
Temperature range (°C)	−20 to 70
Number of cycles	100,000's
Ageing (10 years at RT and nominal voltage)	15% to 30% C loss 40% ESR increase

and to the ensuing power conditioning stage. Moreover, supercapacitors with good power performance can be quite expensive. This can be resolved through tandem use of supercapacitors and batteries in *hybrid storage units*, which can provide energy densities and power densities suitable for the sensor nodes of a WSN. Section 4.5 describes the performance of hybrid storage units used for autonomous sensors.

4.4 Supercapacitor Characterization

This section describes experiments that we performed to determine the parameter values for the circuit model presented in Fig. 4.13(a) when we used it to model several commercial supercapacitors (see Sect. 4.4.2).

4.4.1 Materials and Methods

We measured the R_C of several commercial supercapacitors with an impedance analyzer (HP 4294) at a frequency of 1000 Hz. The modulus of the impedance was considered as R_C because R_{leak} is much higher and the contribution of C was negligible at this frequency. This was also experimentally stated because the phase of the impedance at this frequency was almost zero.

To measure R_C at different temperatures, we used a different method. Supercapacitors were first charged at their maximum operating voltage. Then, a current pulse was drained from them, and the initial voltage drop on the terminals of the supercapacitor was measured. This voltage drop divided by the current amplitude was considered as R_C. We performed several tests for different current amplitude values. The experimental results show the mean value of R_C at the specific temperature. Using this method, we tested four supercapacitors at different temperatures (FCH2730 climatic chamber from CCI).

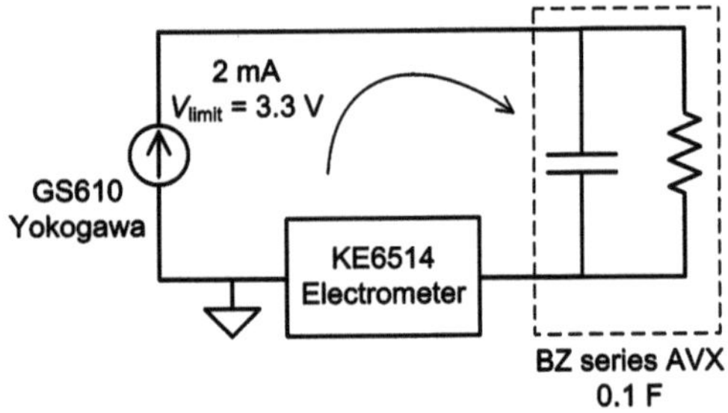

Fig. 4.15 Setup for measuring leakage current in the supercapacitor model

Table 4.8 Main parameter values of the tested supercapacitors, measured at room temperature

Supercapacitor	Capacity (F)	Maximum operating voltage (V)	R_C (Ω) manufacturer (@ 1000 Hz)	R_C (Ω) measured (@ 1000 Hz)
SD series (Panasonic)	0.1	5.5	≤ 75	53
SG series (Panasonic)	0.47	5.5	≤ 30	6.15
SG series (Panasonic)	1	5.5	≤ 30	5.65
HW series (Panasonic)	1	2.3	< 1[a]	0.215
BZ series (AVX)	0.1	5.5	0.096[a]	0.095
BZ series (AVX)	1	5.5	0.050[a]	0.032
PC4 (Maxwell)	4	2.5	0.290[b]	0.120
PC10 (Maxwell)	10	2.5	0.130[b]	0.105

[a]Maximum value.

[b]Typical value.

Leakage current was measured at RT for one of the supercapacitors (BZ series 100 mF, AVX) by using the setup shown in Fig. 4.15. First, the supercapacitor was charged from 0 V to 3.3 V (with the GS610 source, at a constant current of 2 mA), and then, was left floating at 3.3 V for several hours in order to let the slower *RC* branches shown in Fig. 4.13(b) charge completely. Current was measured with the KE6514 electrometer (Keithley) and acquired using a GPIB bus. The measured end-current was considered as leakage. This method is proposed in the electrometer instruction manual [27]. The leakage current was also measured at different temperature levels inside an FCH2730 climatic chamber (CCI).

4.4.2 Experimental Results

4.4.2.1 Series Resistance

The most relevant parameter values of the tested supercapacitors, and the measured values of R_C, are listed in Table 4.8. As observed in the table, the measured values

Table 4.9 R_C values at 0 °C and at 25 °C for four of the tested supercapacitors, measured by applying a discharge pulse to the supercapacitor

Supercapacitor	Temperature (°C)	
	25	0
SG series 0.47 F (Panasonic)	6.76	11.38
BZ series 0.1 F (AVX)	0.113	0.13
PC4 (Maxwell)	0.14	0.223
PC10 (Maxwell)	0.115	0.165

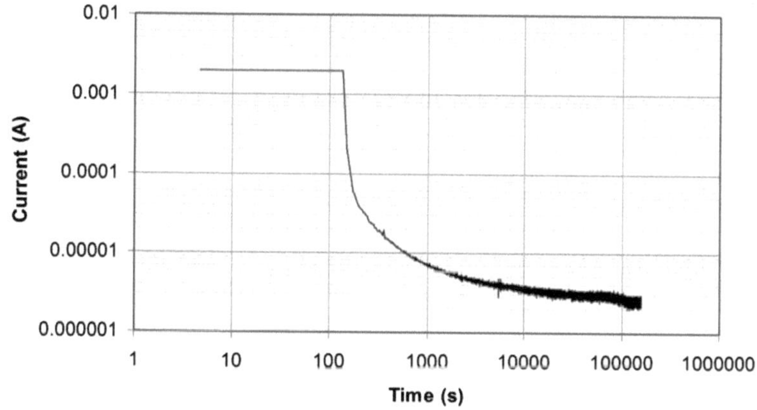

Fig. 4.16 Current delivered by the power source during the charge of the supercapacitor (BZ series 0.1 F, AVX) at room temperature

approach the manufacturer specified values. The R_C value of the first three supercapacitors is very high, because they are intended for back-up applications in which small currents (microampere level) are drained and the voltage drop due to R_C is insignificant.

The R_C values measured at 0 °C and at 25 °C for four supercapacitors are shown in Table 4.9. The values at 25 °C are in good agreement with those presented in Table 4.8. As observed in the table, R_C was higher at the lower temperature.

4.4.2.2 Leakage Current

Figure 4.16 shows (in logarithmic scales) a plot of the current measured while the supercapacitor was being charged, first at constant current and then at constant voltage. As observed in the figure, the current initially had a constant value of 2 mA (constant current charge) and then, once the voltage had reached 3.3 V (*i.e.* the main branch of the model in Fig. 4.13(b) had fully charged), the current steeply decreased. The supercapacitor took several days to completely charge, and the current progressively decreased as more branches (Fig. 4.13(b)) were charged. After 10^5 s, the current delivered by the power source seemed to stabilize at 2.6 μA, which can

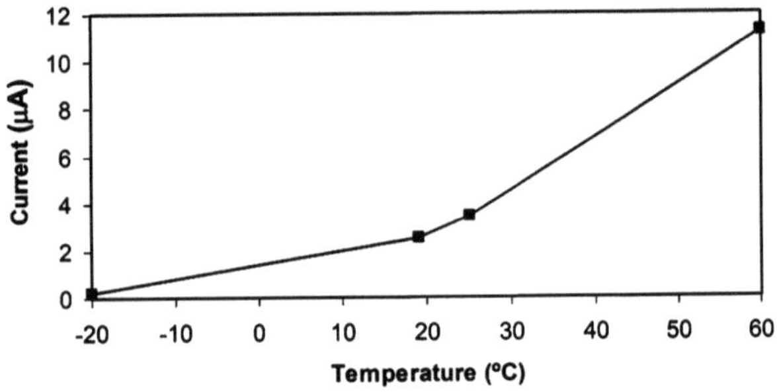

Fig. 4.17 Leakage current at 3.3 V plotted against temperature

be considered as the leakage current (*i.e.* the current drained by R_{leak}). This leakage current is small enough to consider the use of supercapacitors in autonomous sensors.

Leakage was measured at only one voltage; therefore, voltage had to be swept in order to obtain a relationship between capacitor voltage and leakage. Between measurements, some time was needed for the leakage to stabilize.

The leakage current at different temperatures for a voltage of 3.3 V is plotted in Fig. 4.17. As observed in the plot, the leakage current remained less than 12 µA, which is one order of magnitude lower than the leakage current specified by the manufacturer at 70 °C for the full product range (< 100 µA).

4.5 Hybrid Systems

Given[1] that capacitors are complementary to batteries [28, 29] (see Sects. 4.1 and 4.3), they have been harnessed to increase the power capabilities of batteries, even in commercial products [30]. Storage units comprising a battery and a capacitor operating in tandem are known as *hybrid systems*. In [31], the authors used a 600 mAh lithium battery in parallel with a 600 mF supercapacitor to extend the runtime down to a specific threshold voltage under a 2 A GSM pulsed load. The results were only experimentally assessed; no analytical formulation was provided. Dougal et al. [32] theoretically analyzed the power and life extension under pulsed load conditions provided by a hybrid storage unit composed of a supercapacitor in parallel with a battery. However, they did not provide any experimental results and only tackled runtime extension from an energy perspective, disregarding the effect of the voltage drop at the battery terminals. In [33], the researchers used a 1.2 Ah lithium-ion battery in parallel with either a 5 F or a 50 F capacitance storage. The runtime of the hybrid storage unit with the 50 F capacitor was only 5% higher than with the single battery.

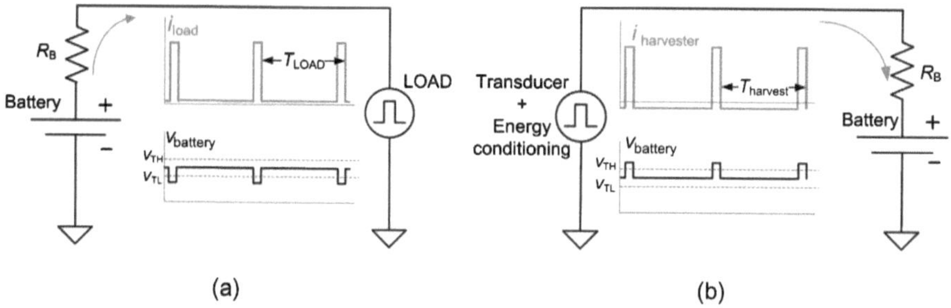

Fig. 4.18 Battery voltage behavior (**a**) under a pulsed load current, and (**b**) when energy is harvested and subsequently delivered as a pulsed current

Despite the extensive work done with hybrid storage units, their advantages have not been thoroughly assessed and tested in low-power wireless sensor nodes, which operate with pulsed load currents on the order of tens of milliamps. Herein we present a comprehensive theoretical analysis that extends literature precedent and provides design guidelines for choosing the best supercapacitor for a given load. The analysis is supported by extensive experimental results (we previously reported this work in [34]). Although we focused on wireless sensor nodes, our analysis and results are applicable to other low-power devices having a pulsed current consumption profile.

4.5.1 Problem Statement

Power to autonomous sensors is provided by primary batteries or by harvesting environmental energy and storing it in secondary (rechargeable) batteries [35]. Choosing a suitable battery is not trivial, owing to the different constraints highlighted in Sect. 4.1. Due to the internal impedance of the battery (Fig. 4.1(b)) and the current profile of autonomous sensors (Fig. 4.2(a)), a voltage drop across the battery terminals appears at the activation time (Fig. 4.2(b)). This phenomenon is illustrated in Fig. 4.18(a). When the battery voltage falls below the minimum supply voltage allowed (V_{TL} in Fig. 4.18(a)), the sensor node stops working, thereby impeding extraction of the available energy from the battery, and consequently, reducing the node runtime [1]. Additionally, some power (and therefore, energy) is lost at the internal impedance of the battery. Small batteries (e.g. button type or thin film lithium batteries) have considerable internal impedance. One way to decrease the internal impedance of batteries is to increase the surface area in which the electrochemical reaction occurs. However, this reduces energy density and increases leakage, and consequently, decreases runtime (Sect. 4.1, [1]). Alternatively, a supercapacitor can be placed in parallel with the battery in order to reduce the equivalent impedance of the storage unit. This strategy is used in some commercial hybrid batteries [30], whereby the current capabilities of bobbin-type thionyl chloride lithium cells are enhanced by adding a hybrid layer capacitor. For this book, we combined generic

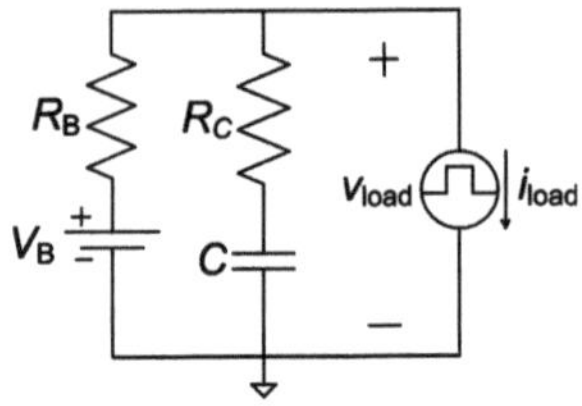

Fig. 4.19 Electrical model of the hybrid storage unit and the sensor node (load)

commercial batteries with supercapacitors, and then tested the resulting hybrid systems as power sources for low-power wireless sensor nodes.

Apart from the pulsed load behavior of autonomous sensors, when energy is harvested from the environment, the current from either the transducer or the energy conditioning stage can also have a pulsed shape (see Sect. 5.3.1). To save energy and boost efficiency at low currents, the energy conditioning stage is not operated constantly; thus, power from the transducer is stored on a capacitor and periodically transferred in high current bursts to the battery. As the current flows into the battery, its voltage increases ($I_o R_B$) (Fig. 4.18(b)), such that I_o (which can be several hundreds of milliamperes) is the peak current injected into the battery during its charge. Secondary batteries generally do not tolerate overcharge well (see Sect. 4.1.4); therefore, designers typically include an over-voltage protection circuit to stop battery charging when an upper voltage limit (V_{TH} in Fig. 4.18(b)) is surpassed. Thus, when the battery operates near this upper limit, or when the product $I_o R_B$ is high, the protection circuit will act too early, and the stored charge will not be as high as expected. Furthermore, some power will be dissipated on R_B. Therefore, hybrid storage units can benefit autonomous sensors by harvesting energy more efficiently and preventing early cuts of the charging current.

For the analyses described in the following sections, we considered the case of a pulsed load. Nonetheless, our methodology could easily be adapted to a pulsed energy source by simply considering a negative I_o and an I_{off} that also accounts for the power consumption of the energy conditioning stage when no energy is transferred to the battery. Finally, when harvesting energy, both situations can be met, and therefore, the capacitor that fulfills the most restrictive condition will probably suffice for both charging and discharging the battery.

4.5.2 Theoretical Analysis

4.5.2.1 Circuit Model

Figure 4.19 shows the electrical model of the hybrid storage unit and the sensor node (load), modeled as a pulsed-current sink (Fig. 4.2(a)). The battery has been modeled with the circuit in Fig. 4.1(a), and the supercapacitor, with the circuit in Fig. 4.13(a). R_{leak} is not shown in Fig. 4.19, but its effect has been considered in the equations as part of the leakage current of the storage unit (I_{leak}). In the remaining part of this section, calculations of the voltage, the power at the load, and the achieved runtime extensions are shown. The expressions for choosing an appropriate supercapacitor for the hybrid storage unit are provided.

4.5.2.2 Load Voltage and Power

Assuming that V_B is constant and the initial voltage at the capacitor is equal to V_B, the load voltage is given by:

$$v_{load}(t) = V_B - V_{off\_leak} - I_o R_B \sum_{n=0}^{N} \left\{ \left[1 - \frac{R_B}{R_B + R_C} e^{-\omega_\beta (t-nT)} \right] \times u(t-nT) - \left[1 - \frac{R_B}{R_B + R_C} e^{-\omega_\beta (t-nT-t_{on})} \right] \times u(t-nT-t_{on}) \right\}, \quad (4.10)$$

whereby $V_{off\_leak} = (I_{off} + I_{leak}) R_B$ is the contribution to the voltage drop of I_{leak} and I_{off}; $u(t)$ is the unit-step function; n is an integer representing the pulse number; and $\omega_\beta = \frac{1}{(R_B+R_C)C}$. Equation (4.10) is similar to that provided in [32], except that it includes the effects of I_{leak} and I_{off}. The load voltage during pulse N tends to:

$$\begin{aligned} v_{load}(NT+t) = {} & V_B - V_{off\_leak} - I_o R_B \left[1 - \frac{R_B}{R_B + R_C} e^{-\omega_\beta t} \left(\frac{1 - e^{-\omega_\beta (T - t_{on})}}{1 - e^{-\omega_\beta T}} \right) \right] \\ & - \left(1 - \frac{R_B}{R_B + R_C} e^{-\omega_\beta t} e^{\omega_\beta t_{on}} \right) \\ & \times \left[u(t - NT - t_{on}) - u(t - NT - T) \right]. \end{aligned} \quad (4.11)$$

This approximation is good enough for $NT > \frac{10}{\omega_\beta}$. At the end of the active time of pulse N, the load voltage is minimal and takes the following value:

$$v_{load}(NT + t_{on}) = V_B - \Delta v_{load}, \quad (4.12)$$

whereby the maximal voltage drop at the internal impedance of the battery (Δv_{load}) is given by:

$$\Delta v_{load} = V_{off\_leak} + I_o R_B \left[1 - \frac{R_B}{R_B + R_C} \frac{e^{-\omega_\beta t_{on}} - e^{-\omega_\beta T}}{1 - e^{-\omega_\beta T}} \right]. \quad (4.13)$$

Alternatively, from (4.11), the average power at the load can be calculated as:

$$\begin{aligned} \bar{P}_{load} & = \frac{1}{T} \int_0^T v_{load}(NT+t) \cdot i_{load} \cdot dt \\ & = \frac{I_o \cdot t_{on}}{T} \left[V_B - V_{off\_leak} - I_o R_B + I_o \frac{R_B^2}{R_B + R_C} \frac{(1 - e^{-\omega_\beta (T - t_{on})})(1 - e^{-\omega_\beta t_{on}})}{\omega_\beta t_{on} (1 - e^{-\omega_\beta T})} \right] \\ & \quad + I_{off} \left(V_B - V_{off\_leak} - I_o R_B \frac{t_{on}}{T} \right) \end{aligned} \quad (4.14)$$

and the average power loss ($\bar{P}_{loss}$) is:

$$\bar{P}_{loss} = \bar{P}_B - \bar{P}_{load}, \quad (4.15)$$

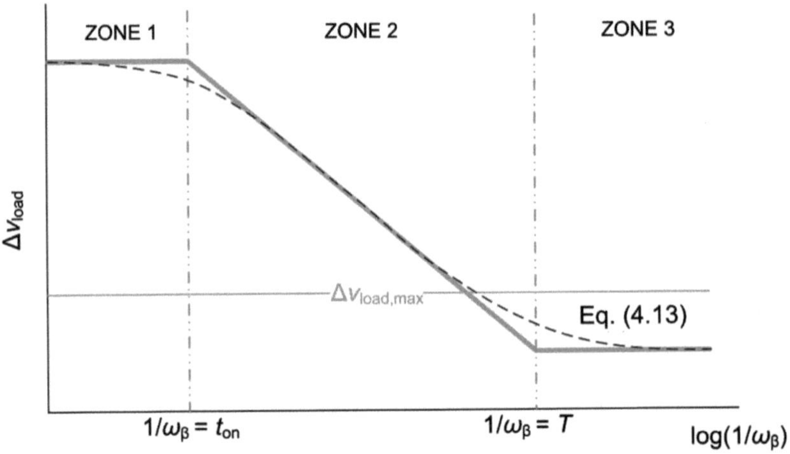

Fig. 4.20 Δv_{load} versus $1/\omega_\beta$. Δv_{load} decreases as C increases

whereby:

$$\bar{P}_{\text{B}} = V_{\text{B}}(I_{\text{o}} \cdot D + I_{\text{leak}} + I_{\text{off}}) \tag{4.16}$$

is the average power provided by the battery.

Assuming that $t_{\text{on}} \ll T$, and considering C (and consequently, ω_β), as a design parameter, (4.13) and (4.14) can be asymptotically evaluated in three zones:

- Zone 1: $\omega_\beta t_{\text{on}} \gg 1$, $\omega_\beta T \gg 1$
- Zone 2: $\omega_\beta t_{\text{on}} \ll 1$, $\omega_\beta T \gg 1$
- Zone 3: $\omega_\beta t_{\text{on}} \ll 1$, $\omega_\beta T \ll 1$

Moving from Zone 1 to Zone 3, Δv_{load} decreases as C increases (and consequently, ω_β decreases).

Evaluating (4.13) in these three zones provides the three asymptotes plotted in Fig. 4.20. The respective expressions are:

$$\begin{aligned}
\text{Zone 1} \rightarrow \Delta v_{\text{load}} &= V_{\text{off_leak}} + I_{\text{o}} R_{\text{B}} \\
&= V_{\text{off_leak}} + I_{\text{o}}(R_{\text{B}} \| R_{\text{C}}) + I_{\text{o}} \frac{R_{\text{B}}^2}{R_{\text{B}} + R_{\text{C}}}, \qquad (4.17\text{a})
\end{aligned}$$

$$\begin{aligned}
\text{Zone 2} \rightarrow \Delta v_{\text{load}} &= V_{\text{off_leak}} + I_{\text{o}}(R_{\text{B}} \| R_{\text{C}}) + I_{\text{o}} \frac{R_{\text{B}}^2}{R_{\text{B}} + R_{\text{C}}} \omega_\beta t_{\text{on}} \\
&= V_{\text{off_leak}} + I_{\text{o}}(R_{\text{B}} \| R_{\text{C}}) + I_{\text{o}} \frac{R_{\text{B}}^2}{R_{\text{B}} + R_{\text{C}}} \omega_\beta D T, \qquad (4.17\text{b})
\end{aligned}$$

$$\text{Zone 3} \rightarrow \Delta v_{\text{load}} = V_{\text{off\_leak}} + I_{\text{o}}(R_{\text{B}} \| R_{\text{C}}) + I_{\text{o}} \frac{R_{\text{B}}^2}{R_{\text{B}} + R_{\text{C}}} D. \tag{4.17c}$$

As an example, Fig. 4.21 shows the load voltage profile at Zone 2, in which the initial voltage drop and Δv_{load} are observed. Moving from Zone 1 to Zone 3, as

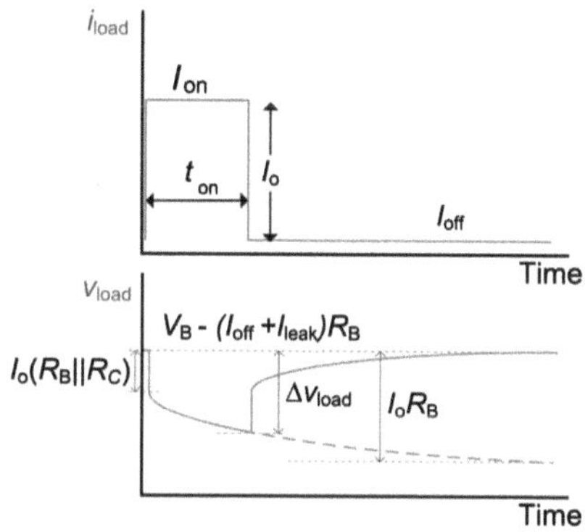

Fig. 4.21 Transient behavior of the load voltage under pulsed load current

C increases, Δv_{load} decreases, as does the peak current provided by the battery. In Zone 1 the hybrid storage unit behaves as a single battery. In Zone 3 the voltage drop is minimal, and it decreases with decreasing values of R_C and D.

From (4.14) to (4.16) the resulting expressions for the power loss at the three zones are:

$$\text{Zone 1} \rightarrow \bar{P}_{loss} = I_o D[V_{off\_leak} + I_o R_B] + P_{off\_leak}, \tag{4.18a}$$

$$\text{Zone 2} \rightarrow \bar{P}_{loss} = I_o D\left[V_{off\_leak} + I_o(R_B \| R_C) + I_o \frac{R_B^2}{R_B + R_C}\frac{t_{on}}{2}\omega_\beta\right] + P_{off\_leak}, \tag{4.18b}$$

$$\text{Zone 3} \rightarrow \bar{P}_{loss} = I_o D\left[V_{off\_leak} + I_o(R_B \| R_C) + I_o \frac{R_B^2}{R_B + R_C} D\right] + P_{off\_leak}, \tag{4.18c}$$

whereby:

$$P_{off\_leak} = I_{off}[V_{off\_leak} + I_o R_B D] + I_{leak} V_B. \tag{4.19}$$

This generates a plot analogous to that of Fig. 4.20; thus, $\bar{P}_{loss}$ is minimal in Zone 3.

4.5.2.3 Runtime Extension

In Sect. 4.5.2.2 V_B was assumed to be constant; however, it actually decreases with Q. Manufacturers usually provide graphs of the battery voltage versus the amount of extracted charge at different discharge rates (*e.g.* Figs. 4.3(b) and 4.4(b)). Thus, the runtime until the battery voltage falls below a specific threshold voltage can be estimated. Runtime increases as either the discharge current or threshold voltage decreases. As a first order approximation and within a determined range, V_B can be assumed to decrease linearly with Q, such that:

$$V_B(t) = V_{B0} - kQ(t) \tag{4.20}$$

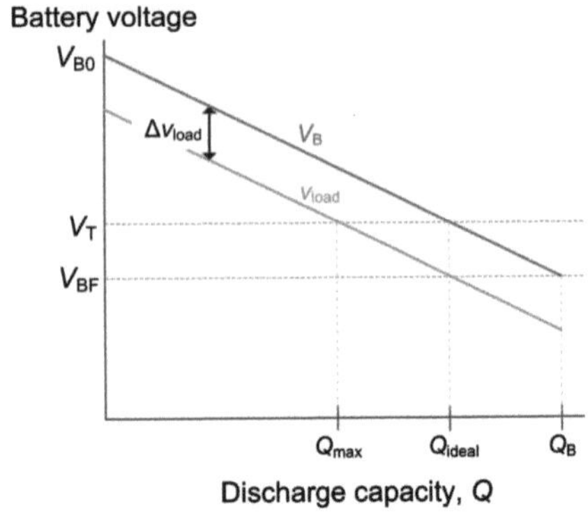

Fig. 4.22 Linear approximation of V_B and v_{load} plotted against discharge capacity

whereby V_{B0} corresponds to the OCV of a fully charged battery; and k (V/As) is a constant that can be inferred from the discharge graphs.

Figure 4.22 represents (4.20) and (4.12) plotted against Q. As observed, the shift between the two plots increases as Δv_{load} increases. Q_{max} is the maximum charge that can be extracted down to a threshold voltage (V_T); Q_{ideal} is the charge that would be extracted at $\Delta v_{load} = 0$; and Q_B corresponds to the total available charge at the battery. V_{BF} corresponds to the battery OCV when there is no charge. When $V_T = V_{BF}$, $Q_{ideal} = Q_B$. The relative extracted charge is given by:

$$\frac{Q_{max}}{Q_{ideal}} = 1 - \frac{\Delta v_{load}}{(V_{B0} - V_T)} \tag{4.21}$$

whereby Δv_{load} can be calculated from (4.13) and the following assumption is made: $V_{B0} > V_T > V_{BF}$. The runtime (t_r) can be estimated as:

$$t_r = \frac{Q_{max}}{DI_o + I_{off} + I_{leak}}. \tag{4.22}$$

Thus, Q_{max} and runtime increase if Δv_{load} is reduced. Runtime also increases with decreasing D. However, for small values of D, the contribution of I_{off} and I_{leak} can be dominant and t_r cannot be increased further. A minimum Δv_{load} (Fig. 4.20, Zone 3) is achieved whenever:

$$C \gg \frac{T}{(R_C + R_B)}. \tag{4.23}$$

From (4.17c), a capacitor with a low R_C further reduces Δv_{load}. A reduction in D, achieved by either reducing t_{on} or increasing T, also reduces Δv_{load}. However, from (4.23), an increase in T leads to a higher value of C, which translates to a larger capacitor and, given the same type and manufacturer, higher cost. Working at Zone 2 (Fig. 4.20), which leads to a smaller value of C, is feasible whenever the achieved Δv_{load} enables the desired charge (Q_{max}) to be obtained from the battery. From (4.21), the maximum permitted voltage drop (Fig. 4.20, $\Delta v_{load,max}$) can be deduced for a desired Q_{max}. Thus, the optimal capacitor (minimum value) can be calculated from (4.17b) as:

$$C > \frac{I_o t_{on} \left(\frac{R_B}{R_B+R_C}\right)^2}{\Delta v_{load,max} - V_{off\_leak} - I_o (R_C \| R_B)} \tag{4.24}$$

which, whenever $R_C \ll R_B$, can be approximated to:

$$C > \frac{I_o t_{on}}{\Delta v_{load,max} - V_{off\_leak} - I_o R_C}. \tag{4.25}$$

As observed in these equations, the required value for C increases with increasing I_o, t_{on} and R_C, and decreasing $\Delta v_{load,max}$.

Whenever the sensor node requires a constant voltage supply (V_{CC}) or the battery voltage does not match the supply voltage range of the sensor node, a voltage regulator (see Chap. 2) must be inserted between the energy storage unit and V_{CC}. Even when a voltage regulator is not strictly needed, its use can increase the overall efficiency. The use of a voltage regulator leads to a constant $\bar{P}_{load}$. Thus, from (4.15), $\bar{P}_B$ will decrease as $\bar{P}_{loss}$ is decreased. Consequently, the runtime will be further extended. From (4.18a)–(4.18c), $\bar{P}_{loss}$ is smaller when a hybrid storage unit is used. $\bar{P}_{loss}$ can be fixed to a small fraction of $\bar{P}_{load}$ in order to reduce the unnecessary power waste at the internal resistance of the battery. Then, a suitable value of C can be chosen from (4.18b), in a similar way as done previously when fixing $\Delta v_{load,max}$.

In conclusion, when high-impedance batteries are equipped with an appropriate capacitor in parallel, their current and power capabilities are enhanced, and the runtime of autonomous sensors can be extended, through reduced internal voltage drop (which implies that more energy can be extracted from the battery) and reduced power loss (which increases the efficiency of the extracted energy transferred to the load).

4.5.3 *Materials and Methods*

We performed accelerated (high D) runtime tests using single batteries and hybrid storage units. In these tests, I_{off} and I_{leak} were much lower than the mean current consumption and did not affect runtime. The source GS610 (Yokogawa) was used as a sink pulsed load (Fig. 4.2) with the following parameter values: $I_{on} = 30$ mA, $I_{off} = 0$ mA, $t_{on} = 100$ ms, and $T = 1$ s. Thus, $I_{c,av} = 3$ mA and $D = 0.1$. The source GS610 was programmed to record the voltage and to switch off its output when the measured voltage fell below a preset voltage.

We also performed tests with a commercial wireless sensor node. A digital multimeter (Agilent 34401) was used to measure the voltage drop at the storage unit. The multimeter was configured with 4.5 digits and high input impedance (> 10 GΩ), and was triggered every 30 s, acquiring 1000 samples in 1.6 s. Figure 4.23 shows the measurement setup.

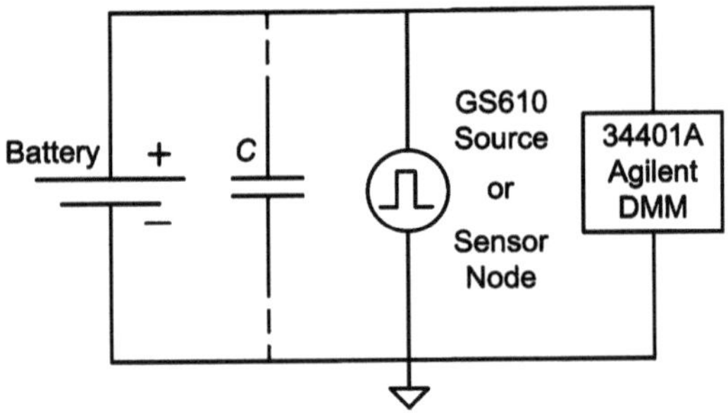

Fig. 4.23 Measurement setup with the storage unit and the load (either a GS610 source or a sensor node). The load voltage was recorded with a multimeter

Table 4.10 Selected batteries and their main characteristics

Battery	Type	Capacity (mAh)	Voltage range (V)	Nominal voltage (V)	R_B (Ω) manufacturer	R_B (Ω) measured	$\Delta v_{load,max}$ (mV)
V357 (Varta)	Primary silver-oxide	165	1.6–1.4	1.55	6[a]	4.79	20
ML 2016 (Maxell)	Secondary lithium	25	3.2–2	3	Not specified	14	80
GP17AAAH2X[b] (GP)	Secondary NiMH	170	2.9–2	2.4	0.24[c]	0.93	150

[a]Not provided by the manufacturer; extracted from [1]. Value measured at 1000 Hz and 100% charge.

[b]Battery pack comprising two GP17AAAH cells in series.

[c]This value was obtained by doubling the typical manufacturer's value for the internal resistance of a GP17AAAH battery at 1000 Hz.

Table 4.10 lists the main characteristics of three selected batteries (one primary and two secondary). R_B was measured as follows: a 100 ms current pulse was drained from the battery (with the GS610 source), and the voltage drop between the battery terminals at the end of the pulse was measured with an oscilloscope (Agilent DSO6032A). The amplitude of the current pulse was tailored to each battery. R_B was then estimated by dividing the voltage drop by the current amplitude. The R_B values in Table 4.10 are on the same order of magnitude as the measured $R_S + R_P$ values listed in Tables 4.4 and 4.5. The maximum discrepancy was exhibited by the lithium battery. This can be explained because different batteries of the same type were used for the two experiments. $\Delta v_{load,max}$ was estimated from discharge curves in order to extract 95% of the total capacity of the battery (Q_B) when voltage fell below the end value of the voltage range (1.4 V for the primary battery and 2 V for the secondary batteries). For the secondary batteries, the manufacturers' discharge curves at the minimum discharge current were used. No discharge curve was available for the primary battery, so an experimental discharge curve was generated from a fresh battery.

Table 4.11 Hybrid storage units

Battery	Charge method	Capacitance (F)	Added capacity (%)
V357	–	1 (BZ series)	0.03
ML 2016	CCCV[a]	0.1 (BZ series)	0.13

[a]Constant-current constant-voltage charge with $I_{\mathrm{limit}} = 2$ mA and 3.2 V. The battery was charged until the current dropped to 30 μA.

The last five supercapacitors in Table 4.8 were considered for their use in the hybrid storage unit. The measured values of R_{C} in these supercapacitors were one to two orders of magnitude lower than the measured values of R_{B} (Table 4.10).

The appropriate value for C in the hybrid storage unit was estimated as follows. Firstly, (4.17a) was compared with $\Delta v_{\mathrm{load,max}}$ for each battery (Table 4.10). The experimental results for R_{B} were used in (4.17a). Whenever the result of (4.17a) is lower than $\Delta v_{\mathrm{load,max}}$, there is no reason to add a capacitor in parallel with the battery; 95% of Q_{B} can be extracted using the battery alone. This was the case for the NiMH secondary battery, which exhibits low internal impedance. Secondly, for each of the remaining batteries, (4.17c) was calculated for the last five supercapacitors in Table 4.8 and subsequently compared with $\Delta v_{\mathrm{load,max}}$. The experimental results for R_{B} and R_{C} were used in (4.17c). When the battery voltage exceeded the supercapacitor maximum operating voltage, two supercapacitors connected in series were considered and R_{C} was multiplied by two. Whenever (4.17c) is higher than $\Delta v_{\mathrm{load,max}}$, the capacitor in question cannot be used. Lastly, to minimize the size of hybrid system, the capacitor having the lowest capacitance that accomplished (4.24) was selected. However, other selection criteria (*e.g.* cost) could be considered. If experimental values for R_{B} and R_{C} are unavailable, then manufacturer's values can be used.

Table 4.11 shows the two resulting hybrid storage units. The secondary battery was recharged using the manufacturer's recommended method. The relative added charge introduced by the corresponding supercapacitor was computed as:

$$Q_{\mathrm{ad}} = \frac{C(V_{\mathrm{B,max}} - V_{\mathrm{B,min}})}{Q_{\mathrm{B}}} \tag{4.26}$$

whereby $V_{\mathrm{B,max}}$ and $V_{\mathrm{B,min}}$ correspond to the maximum and minimum values, respectively, of the battery voltage range in Table 4.10; and Q_{B} is the (charge) capacity of the battery. In both cases, Q_{ad} was well below the nominal capacity of the corresponding battery, and therefore, did not significantly contribute to the runtime extension.

The leakage of the 0.1 F supercapacitor was 2.6 μA at RT and 3.3 V (see Sect. 4.4.2.2). This value is comparable to the sleep currents of commercial transceivers. Active currents of transceivers are roughly 30 mA (Chap. 2); therefore, the leakage currents only dominate when $D < 10^{-4}$. Leakage was smaller at lower voltages and lower temperatures (1.4 μA at 2.5 V; 0.6 μA at 2 V; and 0.2 μA at 3.3 V and −20 °C), but increased at higher temperatures (11.8 μA at 60 °C). Another factor to consider is that leakage in primary batteries may be on the microampere

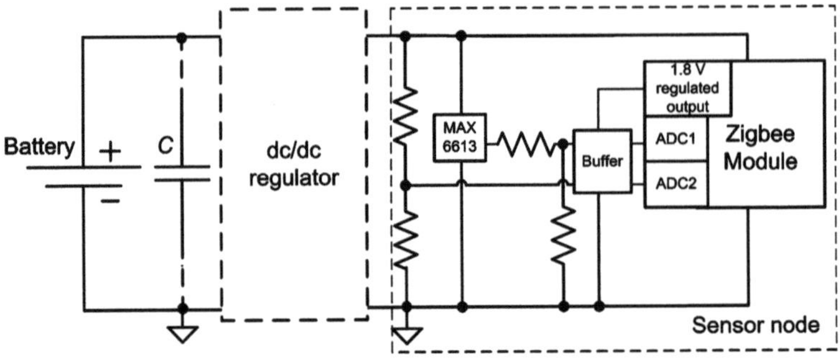

Fig. 4.24 Storage unit and sensor node

level, and it worsens with high temperature. Secondary batteries have even greater leakage (see "Lifetime" in Table 4.1 and "Self-discharge" in Table 4.2).

We performed additional runtime tests using a sensor node as the load and an ML2016 lithium battery (Fig. 4.24). The hybrid storage unit was the same as that used in the previous experiments with this battery (Table 4.11). The sensor node comprised a temperature sensor (MAX6613) and a ZigBee module (ETRX-2 [36], Telegesis). The ZigBee module includes a microcontroller and a transceiver. Recommended voltage supplies for this module range from 2.1 V to 3.6 V. The output of the temperature sensor was matched to the internal 16 b A/D converter of the module with a suitable voltage divider. Supply voltage was also monitored using a high-resistance voltage divider. Buffers were employed to avoid loading effects. Per manufacturers' recommendations, 100 nF decoupling capacitors were used for the temperature sensor and the buffer; however, these capacitors could be left out provided that the supercapacitor, when using the hybrid storage unit, is physically close to the transceiver. The transceiver was programmed to wake up and send a data packet every 1.5 s. Data packets were received by another transceiver node powered from the grid and connected via serial link (EIA-232) to a computer for data processing. The distance between nodes was fixed at 35 cm.

We performed tests at RT and at $-15\,^{\circ}$C with the climate chamber. At RT tests were performed with and without a buck converter (MAX1920) placed between the storage unit and the sensor node. Since the current consumption of the ZigBee module does not vary significantly with supply voltage (see Chap. 2), the dc/dc regulator can help to extend the runtime of the sensor node whenever it is properly used.

4.5.4 Experimental Results

Figure 4.25 shows the discharge profile with the silver-oxide battery and the corresponding hybrid storage unit. The source GS610 was used as load. For the single battery, runtime is strongly dictated by V_{T} and ranges from only 1 h for $V_{\mathrm{T}} = 1.45$ V

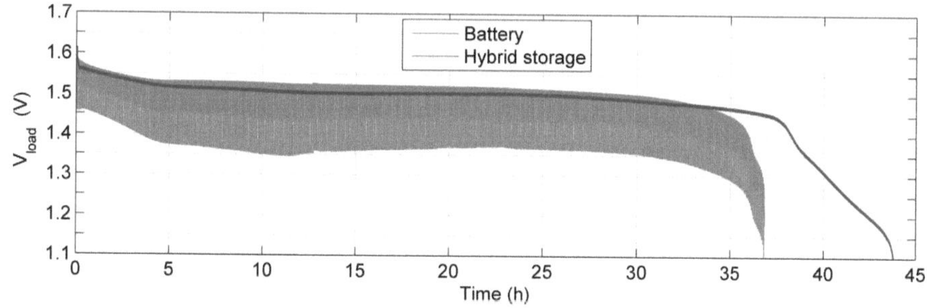

Fig. 4.25 Discharge profile of the silver-oxide battery (*red*) and of the corresponding hybrid storage unit (*blue*). The source GS610 was used as load (color online)

(whereby a large amount of non-profitable charge remains in the battery) to more than 34 h for $V_T = 1.3$ V. The large voltage drop (ranging from roughly 150 mV to 200 mV) is due to the internal impedance of the battery. For the hybrid storage unit, runtime did not change significantly for $V_T < 1.45$ V and was always longer than 37 h. Thus, the addition of a capacitor greatly aids in squeezing the battery. The prior sudden decrease in the battery voltage with the single battery (after 35 h runtime) may have stemmed from slight differences between the two different primary batteries used for the two different storage units.

Figure 4.26 illustrates the results for the lithium battery and the source GS610. For $V_T = 2$ V, the runtimes were 4 h 58 min and 7 h 28 min for the single battery and for the corresponding hybrid storage unit, respectively. The runtime of the hybrid storage unit matches the manufacturer-specified discharge time for a continuous discharge current of 3 mA (7 h 30 min). As can be inferred from the plot corresponding to the single battery (in red), R_B increased with an increasing DOD. In the hybrid storage unit (in blue) the increasing value of R_B did not affect the runtime because the much lower value of R_C dominated the overall contribution to the voltage drop. The two insets show the voltage profile during the active time for the single battery (upper inset) and the hybrid storage unit (lower inset). A much larger voltage drop is observed with the single battery (~ 0.4 V) than with the hybrid storage unit (~ 40 mV).

Figure 4.27 shows the current consumption profile of the sensor node (Fig. 4.24) recorded over 2.5 s with an oscilloscope (Agilent MSO8104A). Network function packets (marked as N) were sent among data packets (marked as D). The current consumption values were 36 mA during data transmission; 9 mA when the sensor is active; and in the microampere level during sleep. The average current consumption, calculated from a 200 s acquisition frame, was 2.23 mA. Figure 4.28 shows the discharge profile of lithium battery and of the corresponding hybrid storage unit when the sensor node was used as load. Temperature data were properly sent until the voltage dropped to 1.65 V (V_T), at which point the ZigBee module stopped working properly. However, this value is far below the minimum voltage recommended by the manufacturer (2.1 V). Runtime for the single battery was 8 h 50 min. Runtime for the hybrid storage unit extended to 10 h 16 min, an increase of 16%. This time value is in accordance with the manufacturer-provided discharge curves

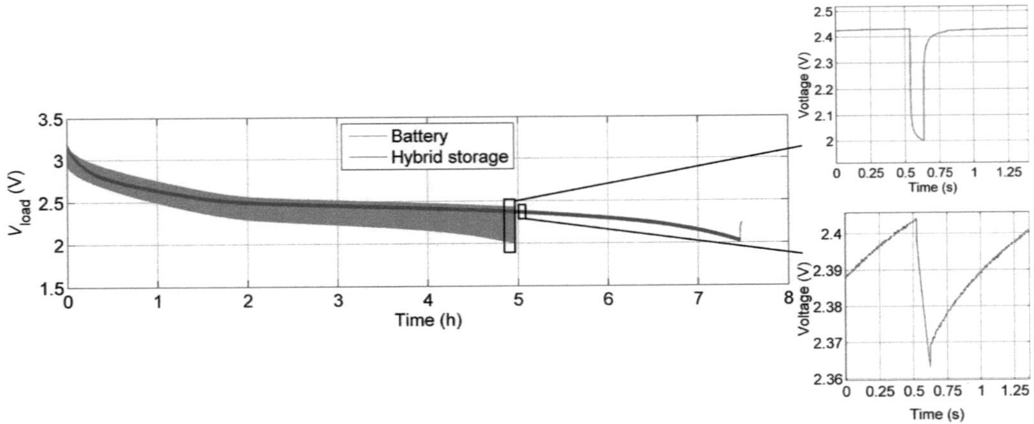

Fig. 4.26 Discharge profile of the lithium battery (*red*) and of the corresponding hybrid storage unit (*blue*). The source GS610 was used as load (color online)

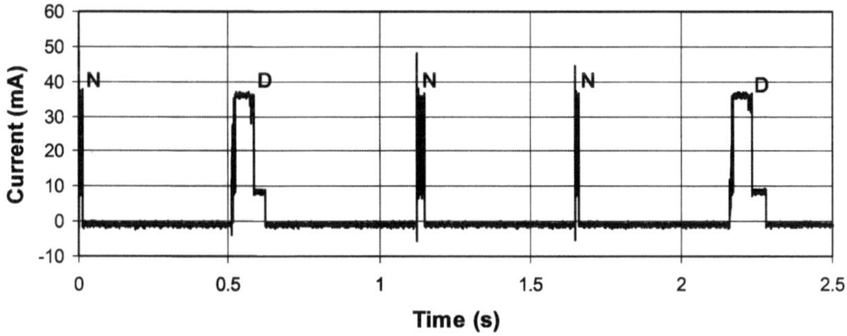

Fig. 4.27 Current consumption profile of the sensor node

for discharge currents between 2 mA and 3 mA. The amplitude of the voltage drops was similar to that shown in Fig. 4.26. In contrast, the node runtime with both storage units was longer than that shown in Fig. 4.26, because of the lower mean current consumption of the load and the lower resulting value for V_T. At $-15\,^{\circ}$C the runtime was extended from 2 h 44 min to 6 h 18 min, an increase of 130% (graphs not shown). Absolute runtimes were lower than those shown in Fig. 4.28, probably due to the lower chemical activity and greater internal resistance of the battery at low temperatures [1].

Figure 4.29 shows an extension of the results presented in Fig. 4.28, illustrating the results of an experiment run with the addition of a MAX1920 buck converter configured to provide a regulated output voltage near 2 V. Voltages at the output of the converter are also represented. As observed in the figure, the converter worked correctly down to an input voltage (from the storage unit) of 2 V. For the single battery the buck converter lengthened the runtime up to this point from 7 h (as shown in Fig. 4.28) to 8 h 8 min. For the hybrid storage unit, it extended the runtime to 10 h 51 min (nearly 2 h longer than that observed in Fig. 4.28) for a $V_T = 2$ V. Thus, the resulting runtime extension in the hybrid storage unit relative to the single battery was 2 h 43 min (a 33% increase), which is longer than when a dc/dc converter was not used (see Sect. 4.5.2.3). Strategic use of the dc/dc converter's shutdown function could reduce the power consumption during the sleep time intervals of the sensor node. This would be feasible, for example, when the transceiver is powered through the converter and the microcontroller is directly powered from the battery providing the shutdown control signal.

4.6 Conclusions

Among the numerous types of primary batteries, lithium-based batteries are the best suited for autonomous sensors: they offer long shelf life, high energy density and high voltage. Nonetheless, they are significantly more expensive and less broadly available than alkaline batteries.

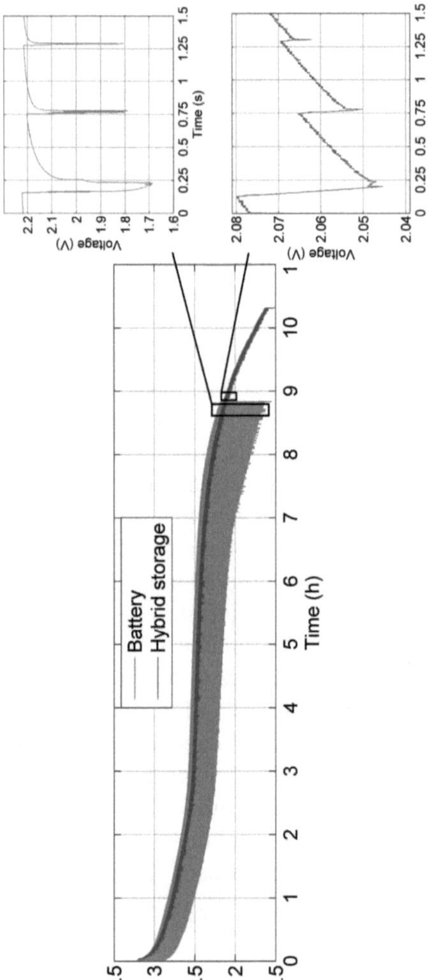

Fig. 4.28 Discharge profile of the lithium battery (*red*) and the corresponding hybrid storage unit (*blue*). The sensor node was used as load (color online)

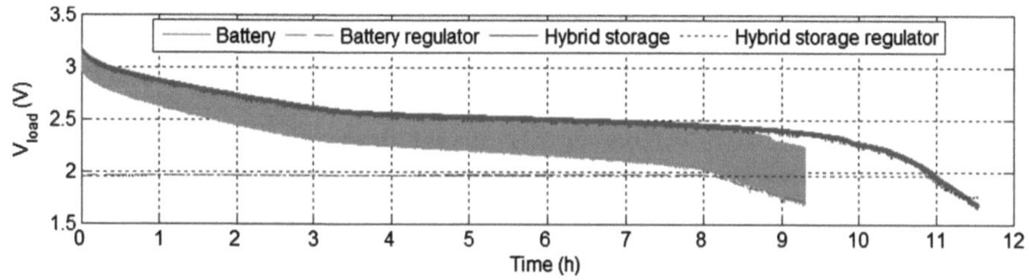

Fig. 4.29 Discharge profile of the lithium battery (*red*) and the corresponding hybrid storage unit (*blue*). A dc/dc buck converter (MAX1920) was added between the storage unit and the sensor node. The voltage at the output of the regulator is also shown (color online)

Available storage units include well-known devices as well as newer devices still in development. Lithium-ion batteries offer higher energy density and lower leakage than other batteries, but present safety issues and generally require a more complicated charging scheme. Furthermore, their voltage can exceed the maximum supply voltage accepted by commercial wireless sensors; therefore, their use may demand a power conditioning stage. Thin film lithium batteries are gaining importance because they offer the best characteristics of Li-ion batteries and are readily integrated into a chip. NiMH batteries are easy to use and cheap, and their charging method can be easily implemented, but they have lower energy density and worse leakage performance. Additionally, they suffer from memory effects, which can reduce their capacity if they are not properly used. Supercapacitors are sometimes preferred for their almost unlimited cycling performance, but their voltage is linearly dependent on the stored charge, and therefore, their use may demand energy and power conditioning stages, which in turn can result in lower total efficiency. Moreover, their energy density is still one to two orders of magnitude lower than that of secondary batteries, and high-performing supercapacitors are much more expensive than secondary batteries.

The best option for a given autonomous sensor depends on the desired optimization parameter (*e.g.* size, cost, or cycle life). When harvesting energy from the environment, the charging current is usually quite low, and consequently, there is minimal need for complex charging and protection circuits. In these cases, simple over or under voltage protection circuits (*e.g.* a comparator plus a disconnection switch) can be considered.

We used a simple model to characterize several types of the most common primary and secondary batteries. The pulsed shape of the current consumption of autonomous sensors, if well characterized, can be used to determine a battery's internal impedance. The shape of the current consumption of autonomous sensors also suggests that OCV could be measured during t_{off}. Nonetheless, the voltage of the battery requires some time (several τ) to return to the OCV. Zinc-air and silver-oxide batteries showed higher internal impedance, which probably stems from their lower capacities and smaller sizes relative to the other batteries tested. The impedance values for the primary batteries were in good agreement with the manufacturers' data (when available). The secondary batteries with lower capacity (and smaller size) showed higher impedance. For the lithium/MnO_2 battery, impedance increased with increasing DOD, as expected. Battery impedance also increased with decreasing temperature. The battery model employed was validated against simulated data. The root mean square error was lower than 3% for almost all the batteries.

We also characterized several commercially available supercapacitors at different voltages and temperatures. Autonomous sensors demand supercapacitors with a series resistance in the milliohm level. Since supercapacitors for back-up applications exhibit series resistances of single or tens of ohms, they are unsuitable for autonomous sensors. The leakage measured for one supercapacitor was in the microampere level, the same order of magnitude as some batteries.

Hybrid storage units comprise a battery and a supercapacitor in parallel. We have used them to extend the runtime of low-power wireless sensor nodes. This first entailed performing a comprehensive theoretical analysis. Sleep and leakage currents were considered, and design guidelines for choosing the appropriate capacitor were established. These guidelines are also useful for selecting an appropriate capacitor when the charge current of the battery comes from an ambient source and is pulsed. Analysis revealed that the supercapacitors with a low series resistance primarily help when using high resistance batteries; consequently, they enable the use of batteries that otherwise would be ruled out *a priori*. Three low capacity batteries (< 200 mAh) were selected: a primary silver-oxide battery, a secondary lithium battery, and a secondary NiMH battery. For the NiMH battery, addition of a supercapacitor did not improve performance because its internal impedance was already low enough. For the remaining two batteries, suitable capacitors were selected. Tests were performed in which an electronic load was used as a pulsed current sink. The hybrid storage units always exhibited longer runtime than the corresponding single batteries.

Lastly, we assessed the performance of a wireless sensor node with the lithium battery. Adding a 0.1 F supercapacitor extended the runtime by 16%, and adding a dc/dc regulator as well, by 33%. Supercapacitor leakage was lower than 3 μA at RT (*i.e.* it was on the same order of magnitude as the sleep current of commercial transceivers). Because active currents of transceivers are roughly 30 mA, the leakage currents only dominate when $D < 10^{-4}$.

References

1. D. Linden, T.B. Reddy, *Handbook of Batteries*, 3rd edn. (McGraw-Hill, New York, 2001)
2. Woodbank Communications, Electropaedia. Battery and energy technologies. Available on: http://www.mpoweruk.com/. Accessed January 22, 2010
3. Duracell, Duracell products. Available on: http://www1.duracell.com/es/products-overview.aspx. Accessed June 2010
4. R.M. Dell, D.A.J. Rand, *Understanding Batteries* (The Royal Society of Chemistry, London, 2001)
5. S. Jacobs, Considerations for primary cell selection. Battery digest (1997, June 2010). Available on: http://www.tadiranbat.com/pdf.php?id=battery_digest_primary_cell_considerations
6. I. Buchmann, *Batteries in a Portable World*, 2nd edn. (Richmond, 2001)

7. PowerStream, Lithium iron phosphate high capacity rechargeable lithium ion batteries. Available on: http://www.powerstream.com/. Accessed March 2010
8. B. Deveney, K. Nechev, T. Guseynov, Large size lithium ion cells based on LiFePO4 cathode material. Available on: http://www.phostechlithium.com/documents/T13-Deveney10thElectrochemicalPowerSources.pdf. Accessed May 6, 2010
9. N.J. Dudney, Thin film batteries for energy harvesting, in *Energy Harvesting Technologies*, ed. by S. Priya, D.J. Inman (Springer Science+Business Media, New York, 2009), pp. 355–363.
10. N.J. Dudney, Solid-state thin-film rechargeable batteries. Materials Science and Engineering B **116**, 245–249 (2005)
11. F. Lasnier, T.G. Ang, *Photovoltaic Engineering Handbook* (Adam Hilger, Bristol, 1990)
12. C. Min, G.A. Rincon-Mora, Accurate electrical battery model capable of predicting runtime and I-V performance. IEEE Trans. on Energy Conversion **21**, 504–511 (2006)
13. E. Barsoukov, J.R. Macdonald, *Impedance Spectroscopy: Theory, Experiment, and Applications*, 2nd edn. (Wiley-Interscience, New Jersey, 2005)
14. A. Mejia-Aguilar, R. Pallàs-Areny, Electrical impedance measurement using voltage/current pulse excitation, in *Proceedings of IMEKO XIX World Congress* (2009), pp. 662–667
15. S. Roundy, D. Steingart, L. Frechette, P. Wright, J. Rabaey, Power sources for wireless sensor networks, in *Proceedings of 1st European Workshop on Wireless Sensor Networks, EWSN* (2004), pp. 1–17
16. I. Buchmann, *Batteries in a Portable World* (Cadex Electronics Inc., Richmond, 2000)
17. Maxwell Technologies Inc, Design considerations for ultracapacitors. Available on: http://www.maxwell.com/pdf/uc/white-papers/200904_WhitePaper_DesignInGuide.pdf. Accessed January 27, 2010
18. Panasonic, Gold capacitors. Technical guide. Available on: http://industrial.panasonic.com/www-ctlg/tech/tABC0000_AM.html. Accessed June 2010
19. J. Schindall, The charge of the ultracapacitors. IEEE Spectrum **44**, 42–46 (2007)
20. J. Miller, P.J. McCleer, M. Cohen, Maxwell Technologies Inc., Energy buffers. Available on: http://www.maxwell.com/pdf/uc/white-papers/200904_WhitePaper_EnergyBuffers.pdf. Accessed June 2010
21. S. Buller, E. Karden, D. Kok, R.W. De Doncker, Modeling the dynamic behavior of supercapacitors using impedance spectroscopy. IEEE Trans. Industry Applications **38**, 1622–1626 (2002)
22. J.N. Marie-Francoise, H. Gualous, A. Berthon, Supercapacitor thermal- and electrical-behaviour modelling using ANN. IEE Proceedings of Electric Power Applications **153**, 255–262 (2006)
23. G. Prophet, Supercaps for supercaches. EDN, Electronic Design Newsletter (2003, June 2010). Available on: http://www.edn.com/article/CA268379.html
24. L. Shinaberger, B. Rawal, C. Reynolds, BestCap®: A new generation of low voltage, low ESR, pulse, double layer capacitors. Available on: http://www.avx.com/docs/techinfo/bestgen.pdf. Accessed January 27, 2010
25. Maxell, Maxell: Product line. Industrial batteries. Available on: http://www.maxell.co.jp/e/products/industrial/battery/index.html. Accessed January 2006
26. P. Mars, C. Greene, RF energy harvesting and battery-free wireless sensors. Available on: http://www.cap-xx.com/resources/docs/CAP-XX&Powercast_RF_Energy.pdf. Accessed January 2010
27. Keithley, Model 6514 system electrometer instruction manual. Available on: http://www.keithley.com/products/dcac/currentvoltage/?path=6514/Documents#6. Accessed January 29, 2010
28. G.A. Rincon-Mora, M. Chen, Squeezing operational life out of a shrinking energy capsule. Power management design line (2006, November 20). Available on: http://www.powermanagementdesignline.com/howto/194500129
29. T. Christen, M.W. Carlen, Theory of ragone plots. Journal of Power Sources **91**, 210–216 (2000)
30. Tadiran Batteries, PulsesPlus batteries for high current pulse applications. Available on: http://www.tadiran.com/index.php/pulses-plus-lithium-cells. Accessed September 2009

31. T.A. Smith, J.P. Mars, G.A. Turner, Using supercapacitors to improve battery performance, in *Proceedings of 33rd Annual IEEE Power Electronics Specialists Conference, PESC* (2002), pp. 124–128
32. R.A. Dougal, S. Liu, R.E. White, Power and life extension of battery-ultracapacitor hybrids. IEEE Trans. Compon. Packag. Technol. **25**, 120–131 (2002)
33. C.E. Holland, J.W. Weidner, R.A. Dougal, R.E. White, Experimental characterization of hybrid power systems under pulse current loads. Journal of Power Sources **109**, 32–37 (2002)
34. M.T. Penella, M. Gasulla, Runtime extension of low-power wireless sensor nodes using hybrid-storage units. IEEE Trans. Instrumentation and Measurement **59**, 857–865 (2010)
35. M.T. Penella, M. Gasulla, A review of commercial energy harvesters for autonomous sensors, in *Proceedings of Instrumentation and Measurement Technology Conference, IMTC* (2007), p. 20
36. Telegesis, Products. Telegesis ZigBee technology. Available on: http://www.telegesis.com/Zigbee/zproducts.htm. Accessed August 2007

Chapter 5
Optical Energy Harvesting

Optical energy harvesting for low-power autonomous sensors exploits the technology developed for high-power solar energy. Indeed, despite the differences in design and power constraints between low-power and high-power applications, most advances in the former have stemmed from work on the latter. Nevertheless, there is still much room for performance improvement, especially regarding power efficiency. Additionally, new methods and circuits designed especially for low-power applications should be proposed and tested.

This chapter begins with a description of the solar cell as an energy transducer, showing a corresponding electrical model and characteristic curves, and explaining how this transducer depends on optical power and temperature. Afterwards, methods and circuits for efficient energy management at different levels of circuit implementation complexity are considered. These include two new methods for extracting maximum power from a solar cell, which we developed, analyzed and tested for low-power applications, and propose here.

Complementary, in [1] we presented a practical study, not included in this book, which compares the use of solar cells to that of primary batteries, provides guidelines for implementing solar-powered autonomous sensors, and shows practical implementations of autonomous sensors for an environmental wireless sensor network (WSN) deployed on our campus.

5.1 Solar Cells

Solar, or *photovoltaic* (*PV*), cells are the most widely used optical transducers. Figure 5.1 shows the equivalent electrical model of a generic silicon solar cell and related curves of both current (I) and power (P) versus voltage (V). The diode models the p-n junction. The parallel or shunt resistance, R_p, models the leakage current at the junction and its value is generally very high. The series resistance, R_s, models the ohmic contacts with the silicon and an increase of its value causes a decrease in the slope of the I–V curve in the region where voltage is nearly constant. Contrariwise, a decrease of R_p causes an increase in the slope of the I–V curve in the

M.T. Penella-López, M. Gasulla-Forner, *Powering Autonomous Sensors*,
DOI 10.1007/978-94-007-1573-8_5,

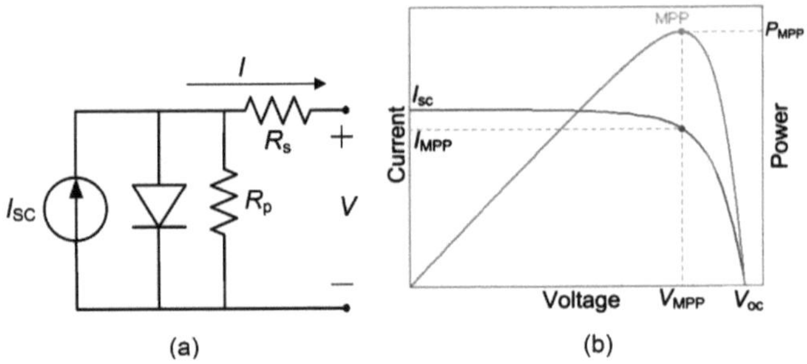

Fig. 5.1 (**a**) Equivalent electrical model, and (**b**) I–V and P–V graphs, of a generic solar cell

region where the current is nearly constant. The short-circuit current, I_{SC}, increases proportionally as the incoming optical power increases. The open circuit voltage, V_{OC}, is roughly constant at high irradiation conditions but varies widely at low irradiations (*i.e.* below 200 W/m^2). I_{SC} remains almost constant with temperature whereas V_{OC} decreases linearly with increasing temperature. P_{MPP} represents the point of maximal generated power (hereafter referred as MPP, for *maximum power point*). I_{MPP} and V_{MPP} are the current and voltage at the MPP, respectively. Power decays more quickly at the right side of V_{MPP} due to the rapid reduction of I in this region.

The relationship between the current and the voltage (I–V curve) of a single solar cell [2] is described by:

$$I = I_{PH} - I_O\left[e^{\left(\frac{q(V+R_S I)}{n_d KT}\right)} - 1\right] - \frac{V + R_S I}{R_p} \tag{5.1}$$

whereby I_{PH} is the photo-generated current, which can be approximated by I_{SC}; I_O is the saturation current of the diode; q is the charge of the electron; n_d is the ideality factor of the diode, which, for silicon, is usually between 1.2 and 1.8 [3]; K is the Boltzmann constant; and T is the cell temperature in Kelvin.

Solar cells of the same material have equal current characteristics per unit area (current density, J). Thus, if the effects of R_p and R_s are ignored for the sake of simplicity, and J is considered instead of I, then (5.1) can be rewritten in a more general form as:

$$J = J_{SC} - J_O\left[e^{\left(\frac{qV}{n_d KT}\right)} - 1\right] \tag{5.2}$$

whereby J_{SC} and J_O are the short-circuit and saturation current densities, respectively. J_O is highly influenced by the material used to construct the cell and by the doping concentrations. Typical values for common materials can be found in photovoltaic handbooks, such as [4]. Nonetheless, considering open circuit conditions ($J = 0$), the relationship between J_O and V_{OC} is obtained by:

$$J_O = \frac{J_{SC}}{e^{\left(\frac{qV_{OC}}{n_d KT}\right)} - 1}. \tag{5.3}$$

Replacing J_O in (5.2) with (5.3) gives the following expression:

$$J = J_{SC}\left(1 - \frac{e^{\left(\frac{qV}{n_d KT}\right)} - 1}{e^{\left(\frac{qV_{OC}}{n_d KT}\right)} - 1}\right), \tag{5.4}$$

which does not depend on J_O. The values of V_{OC} and I_{SC} (and sometimes, of J_{SC}) for a particular cell or PV panel are available on their datasheet.

To obtain the J–V characteristic at different levels of irradiance and at different temperatures, the dependence of J_{SC} and V_{OC} on temperature and irradiance must be considered [4]:

$$J_{SC}(T_{cell}, G) = \frac{G}{1000}\left[J_{SCr} + \frac{dJ_{SC}}{dT}(T_{cell} - T_r)\right], \tag{5.5}$$

$$V_{OC}(T_{cell}, G) = \left[V_{OCr} + \frac{dV_{OC}}{dT}(T_{cell} - T_r)\right]\left[1 + \rho_{OC}\ln\left(\frac{G}{G_{OC}}\right)\ln\left(\frac{G}{G_r}\right)\right], \tag{5.6}$$

whereby T_{cell} is the temperature of the cell (in °C); G is the incident irradiance in W/m^2; dJ_{SC}/dT and dV_{OC}/dT are the current and voltage temperature coefficients, respectively; J_{SCr} and V_{OCr} are the current density and OCV, respectively, at a reference irradiance (G_r) and reference cell temperature(T_r); and ρ_{OC} and G_{OC} (in (5.6)) are two empirical constants used to model the significant variation of V_{OC} at low G. dJ_{SC}/dT is typically very small and dV_{OC}/dT is negative. Typically, $G_r = 1000\ \text{W/m}^2$ ($= 100\ \text{mW/cm}^2$) and $T_r = 25\,°\text{C}$. These reference values are known as *Standard Test Conditions* (STC), because they are the conditions under which manufacturers usually test PV panels. Values of $\rho_{OC} = -0.04$ and $G_{OC} = 1000\ \text{W/m}^2$ are adequate for many silicon PV modules [4]. When directly illuminated, solar cells heat up to above the ambient temperature (T_a). Thus, T_{cell} can be obtained from an empirical parameter known as the *Nominal Operating Cell Temperature* (*NOCT*):

$$T_{cell} = T_a + \frac{NOCT - 20}{800\ \text{W/m}^2}G. \tag{5.7}$$

NOCT is the temperature of the cell when exposed to $800\ \text{W/m}^2$ at $T_a = 20\,°\text{C}$ and wind speed of 1 m/s. It is empirically determined, and for silicon solar cells ranges between 42 °C and 48 °C.

We employed (5.4) to (5.7) to model a single cell of arbitrary area. The parameter values were taken from typical industrial monocrystalline solar cells [4]: $J_{SCr} = 35\ \text{mA/cm}^2$; $V_{OCr} = 0.6\ \text{V}$, $dJ_{SC}/dT = 12.5\ \mu\text{A/cm}^2/°\text{C}$; $dV_{OC}/dT = -2\ \text{mV/°C}$; $NOCT = 42\,°\text{C}$; and $n_d = 1.5$. The computed J–V and S–V curves at several values of G and at $T_a = 25\,°\text{C}$ are shown in Fig. 5.2 (in which S represents the power density of the cell). A curve joining the MPPs (S_{MPP}) is also shown. At high irradiance levels, V_{OC} slightly decreases because of the increase in T_{cell} (due to (5.6) and (5.7)). However, V_{OC} significantly changes at low values of G. These effects can also be appreciated in the curve joining the MPPs.

Figure 5.3 shows the computed J–V and S–V curves for a single solar cell at several values of T_a and at $G = 1000\ \text{W/m}^2$. As can be seen, both V_{OC} and S_{MPP} decrease as temperature increases, whereas J_{SC} barely varies with temperature.

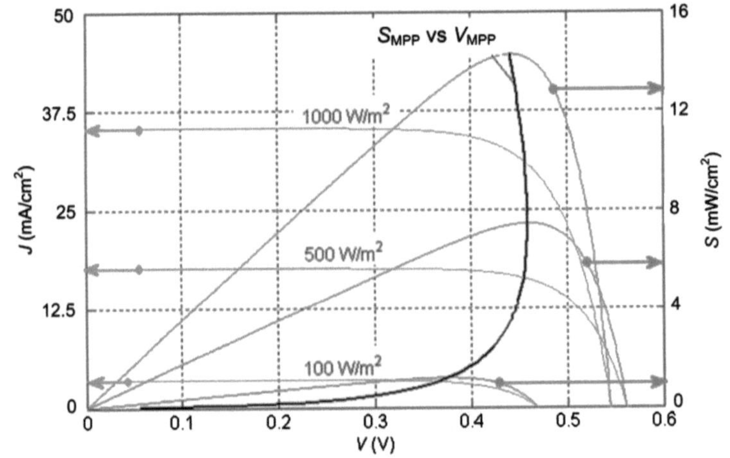

Fig. 5.2 Generic J–V and S–V plots at several values of G and at $T_a = 25\,^\circ\text{C}$ for a single solar cell. A curve joining the MPPs is also plotted

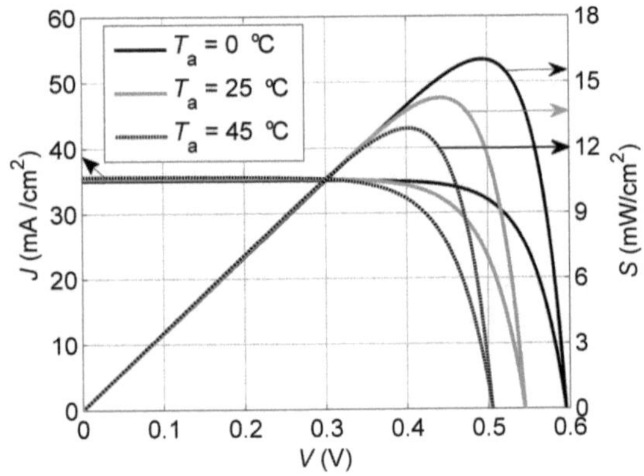

Fig. 5.3 Generic J–V and S–V plots at $G = 1000\ \text{W/m}^2$ and at several values of T_a for a single solar cell

Figure 5.4 shows the computed variation of the relative PV power density (S/S_{MPP}; expressed as percentage) plotted against voltage near V_{MPP} at several values of G. At very low irradiance levels ($10\ \text{W/m}^2$), a small voltage deviation from V_{MPP} ($\sim \pm 12$ mV) causes S to decrease to below 90% S_{MPP}. This is mainly due to the great reduction in V_{MPP} at very low irradiance levels (see Fig. 5.2). At irradiance levels over $50\ \text{W/m}^2$, the voltage deviation that produces the same reduction in S/S_{MPP} is quite higher and is steady. Figure 5.4 also reveals that power density varies more steeply when $V > V_{MPP}$ than when $V < V_{MPP}$.

S/S_{MPP} versus V around V_{MPP} at $G = 1000\ \text{W/m}^2$ was also computed at several values of T_a (Fig. 5.5(a)) and n_d (Fig. 5.5(b)). As observed in the figures, only minor variations were produced. For the remaining computations shown in this chapter, we used an intermediate value of $n_d = 1.5$.

So far in this chapter, only graphs with the current density of single solar cells have been shown. To obtain the I–V characteristic of a single cell of area A, current

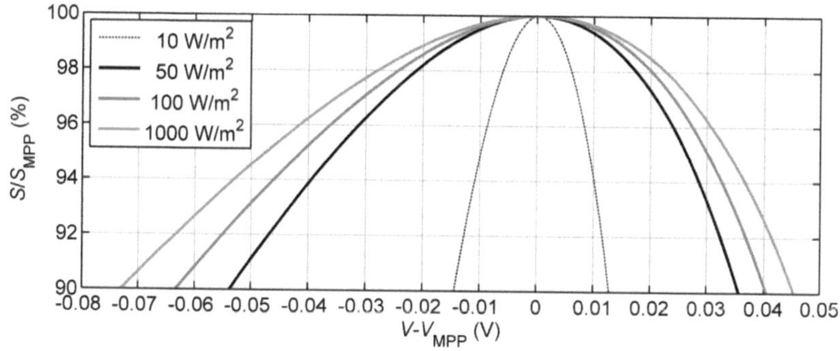

Fig. 5.4 Power density variation near the MPP

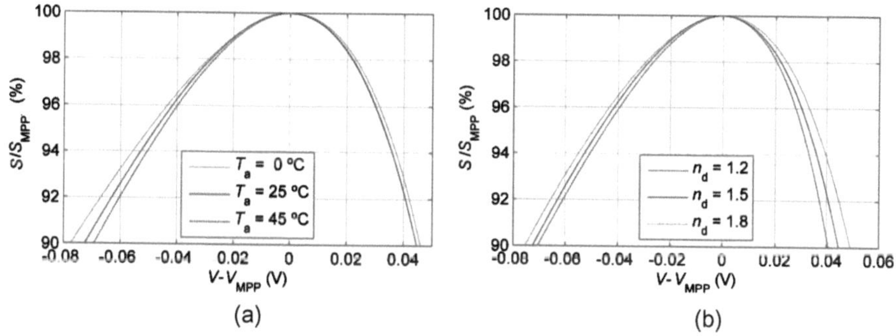

Fig. 5.5 Power density variation around the MPP at $G = 1000\ \mathrm{W/m^2}$ and at three values of: (**a**) T_a or (**b**) n_d

density in (5.4) must be multiplied by A. The interconnection of N cells in series and M cells in parallel within a PV module has an I–V characteristic that can be derived by scaling up the I–V characteristic of one of the cells by factors of N in voltage and M in current, provided that the cells are identical.

5.2 PV Array Simulator

To perform robust tests on the energy conditioning circuits described in the ensuing sections, the I–V curve of the solar cell or panel in question must be stable and reproducible. Commercial PV array simulators are used for high-power PV panels but are unsuitable for the low-power applications that we sought ($<$ 1 W). Consequently, we implemented an *ad hoc* PV array simulator, connecting an accurate current source (GS610, Yokogawa) for the required current level in parallel with a commercial PV panel coated with an opaque cover (Fig. 5.6). The current source simulates the I_{SC} of the PV panel. Provided that the value of R_s of the PV panel is low enough, this solution generates an appropriate I–V curve. Since the panel is not illuminated, $NOCT = 20\,°\mathrm{C}$ (*i.e.* the cell is not overheated).

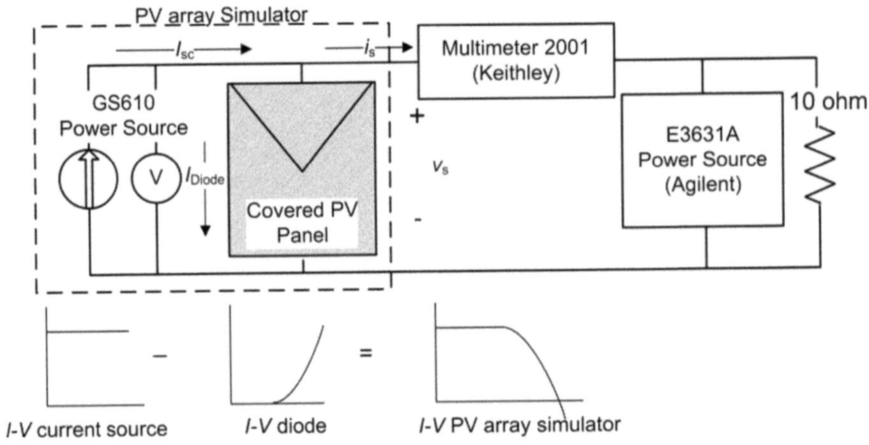

Fig. 5.6 The PV array simulator, and the setup used for its characterization

Fig. 5.7 *I–V* and *P–V* curves of the PV array simulator at 24 °C, and a curve joining the MPPs

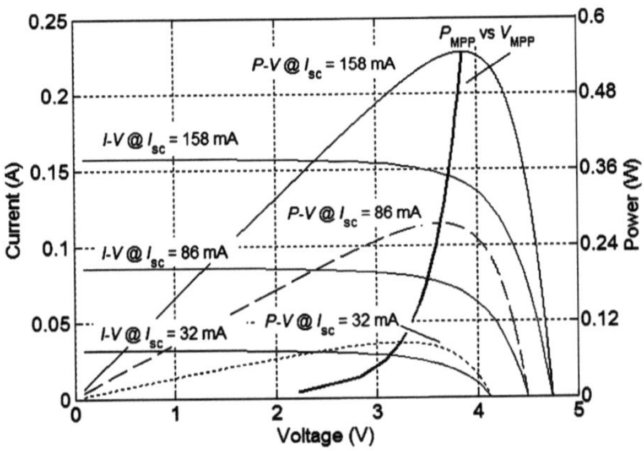

For the PV panel (MSX-005, from BecoSolar) [5] that we used to test energy conditioning circuits (Sects. 5.3.2, 5.5.2 and 5.6.6), $V_{OC} = 4.6$ V, $I_{SC} = 160$ mA and $P_{MPP} = 500$ mW at STC. The voltage temperature coefficient of the panel is $dV_{OC}/dT = -16$ mV/°C, which implies $N = 8$.

The current source was configured to source a constant current (I_{SC}) that would cover the full range of the PV panel, varying from 5 mA to 158 mA in 9 mA steps. For the I_{SC} range considered, the PV array simulator was characterized by using the GS610's measurement unit to measure the panel's voltage; a 2001 multimeter (Keithley) to measure the current of the panel; and a programmable voltage source (Agilent E3631A) in parallel with a 10 Ω/1 W resistor as load (Fig. 5.6). All the instruments were controlled using the program LabVIEW via the GPIB bus. For each current value (I_{SC}), the voltage of the E3631A was increased from 0 V to 5 V in 0.1 V steps. The output voltages (v_s) and currents (i_s) of the PV array simulator were then measured, and the power values, calculated. The *I–V* and *P–V* curves and the MPP points were obtained using a cubic spline interpolation. Figure 5.7

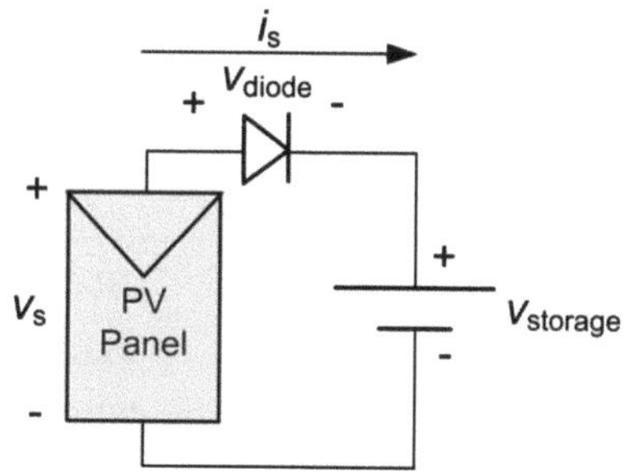

Fig. 5.8 Block diagram of a direct-coupled circuit for a PV panel

shows a set of measured $I-V$ curves and the corresponding $P-V$ curves for the PV array simulator at roughly 24 °C, plus a curve joining the MPPs. Thus, each generated I_{SC} was correlated to a resulting P_{MPP}. The end values for P_{MPP} were 8.2 mW and 545.9 mW, and for I_{MPP}, 3.7 mA and 141.4 mA. Figure 5.7 clearly shows that the PV panel was not overheated.

5.3 Direct-Coupled Circuits

The simplest energy conditioning circuits for PV panels, known as *direct-coupled* circuits, use a blocking diode to avoid the discharge of the storage unit during periods of darkness [6–8] (Fig. 5.8). The working point of the PV panel (v_s) is fixed by the diode voltage drop (v_{diode}) plus the storage unit voltage ($v_{storage}$). Schottky diodes are employed to obtain a lower v_{diode}. Diode losses can be further reduced using other methods: for example, with a switch that bypasses the diode during charging of the storage unit [9]. Proper choice of PV panel and battery can yield a system operating near the MPP, as explained in Sect. 5.3.1. However, V_{MPP} depends on irradiance and temperature; consequently, large variations are expected outdoors. Furthermore, v_{diode} and $v_{storage}$ also depend on temperature and irradiance (i_s). Additionally, $v_{storage}$ changes with the stored charge (see Chap. 4), particularly in supercapacitors. High impedance batteries also present significant voltage variations with the charging or discharging current. Therefore, the system will occasionally work at the MPP.

Some authors (see [10, 11] or [12]) have proposed using a dc/dc converter between the PV panel and the storage unit. The input voltage of the converter (*i.e.* the output voltage of the PV panel) is fixed and the output voltage follows $v_{storage}$. For the system to work near the MPP, the input voltage of the converter must be chosen after thorough characterization of the PV panel. The activation or deactivation of the converter can be controlled by software, to avoid the discharge of the storage unit during periods of darkness. The circuit's efficiency is dictated by the selected dc/dc converter, which enables use of storage units and PV panels with different voltage ranges and is immune to changes in $v_{storage}$. Nevertheless, fixing the output voltage of the PV panel cannot guarantee that the system will operate at the MPP for different values of G and T_a.

5.3.1 Analysis

The direct-coupled circuit's performance can be evaluated according to its total efficiency (η_T):

$$\eta_T \equiv \frac{P_{out}}{P_{MPP}} = \frac{P_s(v_s) - P_{loss,diode}}{P_{MPP}} \tag{5.8}$$

whereby P_{out} is the power delivered to the load (storage unit); $P_s(v_s)$ is the power that the panel delivers at v_s; and $P_{loss,diode}$ is the power dissipated on the diode. Note that η_T accounts for the power loss in the diode and the power loss caused by deviation from the MPP.

We simulated the direct-coupled circuit shown in Fig. 5.8 and computed η_T for different scenarios. For the PV panel, an array of N single solar cells (1 cm^2 area each) in series was considered. For each cell the model described in Sect. 5.1 was employed, using the parameters with which we generated the curves shown in Fig. 5.2. For the Schottky diode, we used a BAT47 (ST Microelectronics). Only batteries were considered as storage units. v_{diode} is typically around 0.4 V at currents in the milliampere level. Thus, to keep the relative diode losses ($P_{loss,diode}/P_{MPP}$) below 20%, storage units with $v_{storage} > 2$ V were chosen.

One strategy for minimizing the effects of diode losses is to run both the PV panel and the battery at high voltages. Nevertheless, when connecting several batteries in series, the charge distribution must be carefully controlled. Furthermore, if the voltage of the storage unit is well above the working voltage of the load (2 V to 3.6 V), a step-down dc/dc converter must be used between the battery and the load. The efficiency of the converter depends on the ratio of the input voltage to the output voltage (V_{IN}/V_{OUT}). With step-down converters, efficiency decreases as V_{IN}/V_{OUT} increases [13]. Thus, we restricted the simulations to storage units with either one Li-ion battery, two NiMH batteries in series, or three NiMH batteries in series. Nonetheless, the NiMH associations must consider charge distribution. Chapter 4 described these two types of batteries and gave their expected voltages. The expected voltages when charging the batteries are: 3.6 V $< v_{storage} <$ 4.2 V for one Li-ion battery; 2.4 V $< v_{storage} <$ 3 V for two NiMH batteries; and 3.6 V $< v_{storage} <$ 4.5 V for three NiMH batteries. Regardless of the voltage range, a step-down regulator can make the load work at its minimum voltage supply, thereby reducing its power consumption (see Chap. 2).

Figure 5.9(a) shows η_T for several values of N for the case in which $v_{storage}$ simulates a Li-ion battery as the storage unit ($v_{storage} = 3.9$ V). When $N = 8$, $v_s(v_{diode} + v_{storage})$ is well above V_{MPP} of the PV panel; consequently, η_T adopts poor values ($< 60\%$). η_T reaches its maximum value at $G = ca.$ 600 W/m^2. Panels with $N = 9$ exhibit high efficiency (*ca.* 90%), mainly limited by the power loss at the diode, for $G > 200$ W/m^2. At lower irradiances, η_T steeply decreases. For $N > 9$, η_T sharply increases until a maximum (*ca.* 90%) at low values of G; decreases for intermediate values of G; and then subtly increases again for the highest values of G, due to the self-heating effect in the PV panel. For $N > 10$, η_T drops at medium to high values of G, and does not markedly increase at low values of G.

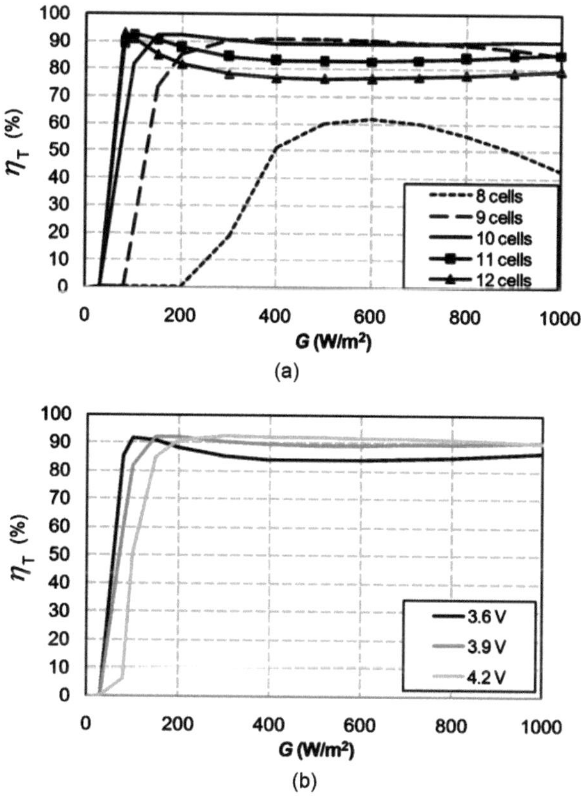

Fig. 5.9 η_T of a direct-coupled system (**a**) for several values of N at $v_{storage} = 3.9$ V and (**b**) for $N = 10$ at several values of $v_{storage}$. The voltages are within the range of those for a Li-ion battery

Panels with $N = 10$ offer the best tradeoff in terms of η_T across the full range of G. Figure 5.9(b) shows η_T for the case of $N = 10$ and $v_{storage} = 3.6$ V, 3.9 V and 4.2 V. As observed in the figure, at high values of G, the best η_T is achieved at maximum $v_{storage}$(*i.e.* mostly charged battery), whereas at low values of G, the best η_T is achieved at minimum $v_{storage}$ (*i.e.* mostly discharged battery). This can prove advantageous for field operation, since on sunny days (*i.e.* high G) the battery will be mostly charged, whereas on cloudy days (*i.e.* low G) it will be mostly discharged.

Figure 5.10 presents the case in which $v_{storage}$ simulates a storage unit with three NiMH batteries connected in series. The best option for the PV panel is again $N = 10$ (Fig. 5.10(a)), just as in the case of Fig. 5.9(a); this is coherent, given that the corresponding $v_{storage}$ values are very close: 4.05 V for the former and 3.9 V for the latter. However, the voltage difference between the minimum and maximum voltages of $v_{storage}$ (0.9 V; see Fig. 5.10(b)) is higher than in the case of Fig. 5.9(b) (0.6 V), which leads to greater dispersion of η_T values, particularly at low values of G.

Figure 5.11 shows the case in which $v_{storage}$ simulates a storage unit with 2 NiMH batteries connected in series. As observed in the figure, for 2 NiMH batteries PV panels with $N = 7$ (Fig. 5.11(a), $v_{storage} = 2.7$ V) seem to be the best

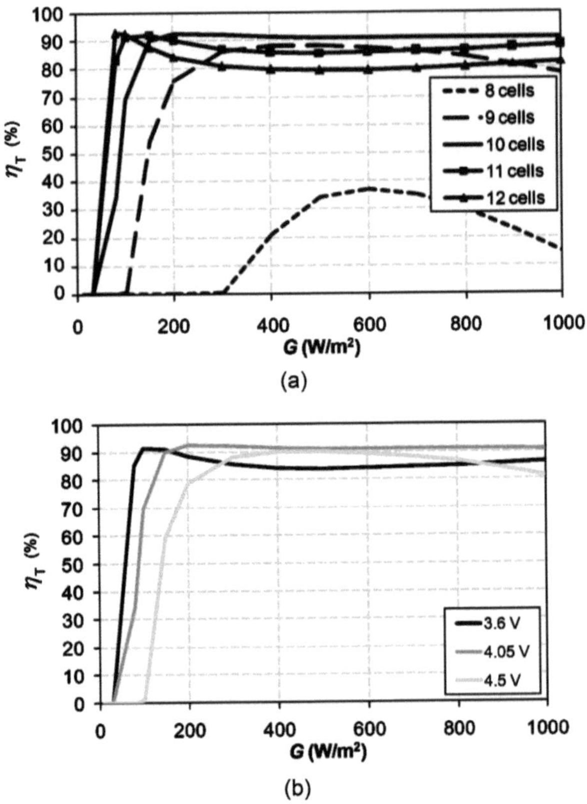

Fig. 5.10 η_T of a direct-coupled system (**a**) for several values of N at $v_{storage} = 4.05$ V and (**b**) for $N = 10$ at several values of $v_{storage}$. The voltages are within the range for three NiMH batteries connected in series

option, corresponding to three single solar cells less than the case of one Li-ion battery (Fig. 5.9(a)). This difference is due to the reduction of $v_{storage}$ (-1.2 V). The efficiencies are slightly lower than those seen for one Li-ion battery (Fig. 5.9(a)) because of the higher relative $P_{loss,diode}$. Figure 5.11(b) shows η_T for different values of $v_{storage}$, which correspond to the expected voltage range of the storage unit.

Figure 5.12 illustrates the results of the simulations performed at several values of T_a, $N = 10$, and $v_{storage} = 3.9$ V (Li-ion battery), which can emulate operational fluctuations such as seasonal effects on η_T. At medium to high irradiance values, η_T increases with increasing T_a; however, higher T_a translates to less available power at the MPP for the same irradiation level (Fig. 5.3). Contrariwise, at low irradiance levels, η_T is highest at $T_a = 0\,°\mathrm{C}$; furthermore, lower temperature also provides more power at the MPP. A field example of this scenario is operation on cloudy winter days.

The best tradeoff among low and high irradiances seems to be achieved at $T_a = 25\,°\mathrm{C}$. As V_{OC} and (consequently, V_{MPP}) changes with temperature ($dV_{OC}/dT = -2$ mV/°C, for a single cell), the optimum N of the PV panel in question will depend on the working temperature and on the design criteria. For example, a design indicated for outdoors, where temperature and available sunlight are dictated by

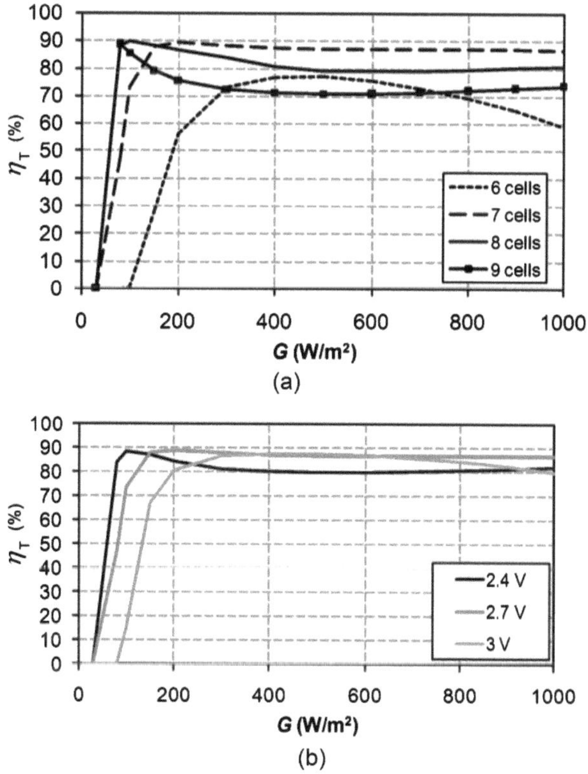

Fig. 5.11 η_T of a direct-coupled system (**a**) for several values of N at $v_{storage} = 2.7$ V and (**b**) for $N = 7$ at several values of $v_{storage}$. The voltages are within the range for two NiMH batteries connected in series

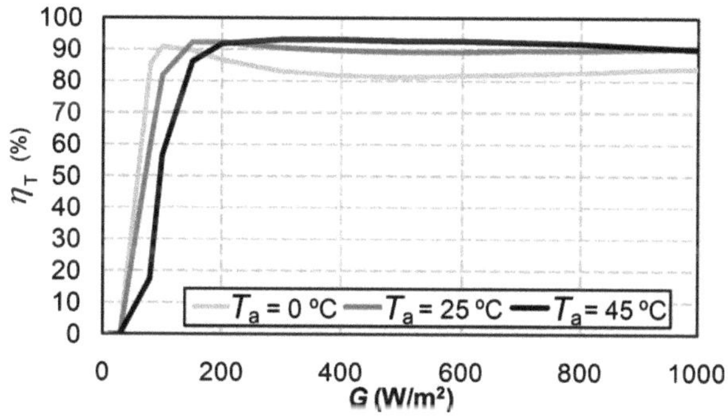

Fig. 5.12 η_T of a direct-coupled system for different T_a and $N = 10$ at $v_{storage} = 3.9$ V (Li-ion battery)

geography, could be optimized for performance in winter, when less solar hours are available. Alternatively, a panel operating indoors in a climate-controlled building may not require seasonal or geographic adjustments, as temperature and irradiance levels may be relatively constant.

Replacing the diode with a switch that disconnects the battery in the absence of light (as in [9]) obviates the diode losses, but adds switch losses instead. Moreover, the switch must be controlled, which implies more complex circuitry and more wasted power. If the added power waste is lower than $P_{loss,diode}$, then the switch sys-

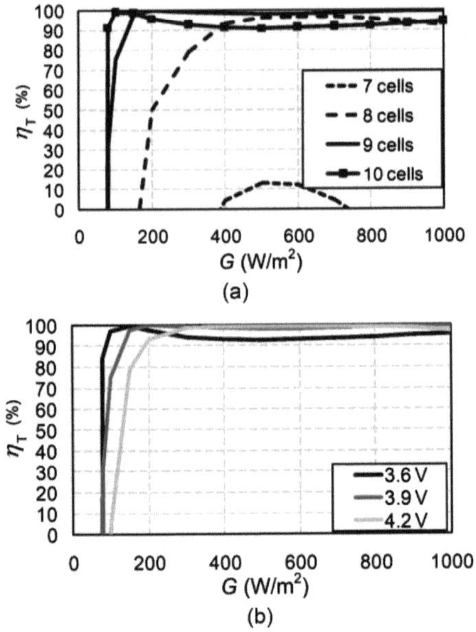

Fig. 5.13 η_T of a system equipped with a switch instead of a diode (the switch losses and control losses are not considered): (**a**) at several values of N and $v_{storage} = 3.9$ V and (**b**) at $N = 9$ and at several values of $v_{storage}$. The voltages are within the range for a Li-ion battery

tem could be preferable to the direct-coupled one. Figure 5.13(a) shows η_T values for a switch system at different values of N and $v_{storage} = 3.9$ V (Li-ion battery), and without considering the switch and control power losses. The maximum η_T approaches 100% because there is no diode, and consequently, $P_{loss,diode} = 0$ in (5.8). The best value in this case appears to be $N = 9$, which is one cell less than when using the diode (see Fig. 5.9(a)). These values are consistent, as v_{diode} nearly matches the V_{MPP} of one single cell at most irradiance values (Fig. 5.2). Figure 5.13(b) shows η_T at $N = 9$ and for the voltage range of a Li-ion battery. At high irradiances, η_T is always greater than 90%. Similar conclusions to those reached with Fig. 5.9(b) can be drawn with Fig. 5.13(b).

We have explained how to choose a PV panel with the suitable number of single solar cells connected in series in order to obtain the best results with a direct-coupled circuit, and considering different types and combinations of rechargeable batteries. Nevertheless, to obtain the desired output power, designers must still size the PV module. Therefore, they must consider that an increase (or decrease) in the panel area will lead to higher (or lower) currents, and consequently, will generate a small increase (or decrease) in relative diode losses (represented by the ratio $P_{loss,diode}/P_{MPP}$) because of the slightly higher (or lower) value of v_{diode}.

5.3.2 Experimental Results

We performed laboratory tests and field experiments to assess the performance of a direct-coupled circuit for powering autonomous sensors.

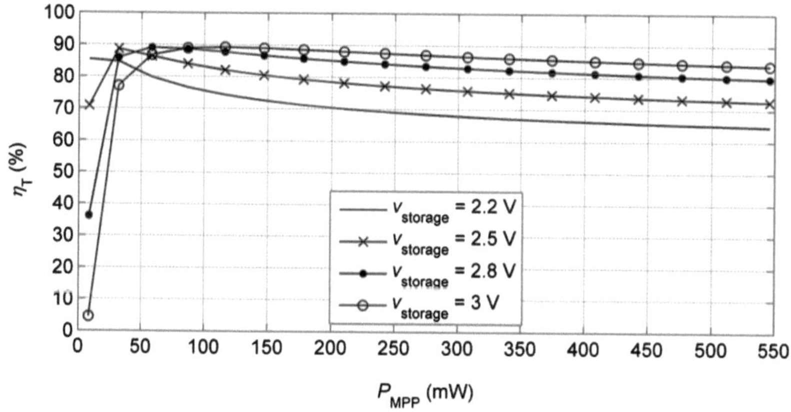

Fig. 5.14 Measured η_T at different values of $v_{storage}$ and P_{MPP}. A modified version of the setup corresponding to Fig. 5.6 was used, which incorporated a BAT47 Schottky diode inserted at the output of the PV array simulator

5.3.2.1 Laboratory Measurements

For the laboratory experiments, a modified version of the setup corresponding to Fig. 5.6 was used, in which a BAT47 Schottky diode was inserted at the output of the PV array simulator. η_T was experimentally measured for an output voltage range, fixed by the E3631A voltage source, from 2.2 V to 3.0 V, thus emulating a storage unit of two NiMH batteries connected in series. For each voltage value, the GS610 current source generated the whole I_{SC} range. The results, illustrated in Fig. 5.14, are similar to the computed data shown in Fig. 5.11(b), but differ because $N = 8$, $NOCT = 20\,°C$, and the single solar cells of the PV panel do not have an area of 1 cm^2. Figure 5.15 shows computed data that account for the parameter values of the PV panel (extracted from the manufacturer's data [5]): $A = 4.57$ cm^2; $J_{SCr} = 35$ mA/cm^2; $V_{OCr} = 0.575$ V; $dJ_{SC}/dT = 32.8$ μA/cm^2/°C; $dV_{OC}/dT = -2$ mV/°C; $N = 8$; $NOCT = 20\,°C$; and $n_d = 1.5$. The value of A was obtained by dividing I_{SC} at STC (160 mA) by J_{SCr}. V_{OCr} was computed by dividing the V_{OC} at STC (4.6 V) by N. The simulated and experimental data are generally in good agreement at high irradiance values; the greatest discrepancy occurs below 100 W/m^2 ($P_{MPP} < 50$ mW in Fig. 5.14). Maximum efficiencies were roughly 90%, suggesting that the diode loss was approximately 10%.

5.3.2.2 Field Measurements

We performed field measurements using the circuit shown in Fig. 5.8, equipping it with two NiMH batteries connected in series as the storage unit. The batteries were initially in a discharged state. To determine how η_T varies over the course of a day, the module was placed under direct sunlight on a sunny afternoon in September (maximum T_a: *ca.* 27 °C). The current delivered to the storage unit and its voltage were measured ($P_{storage}$). The PV panel was alternatively connected to a voltage

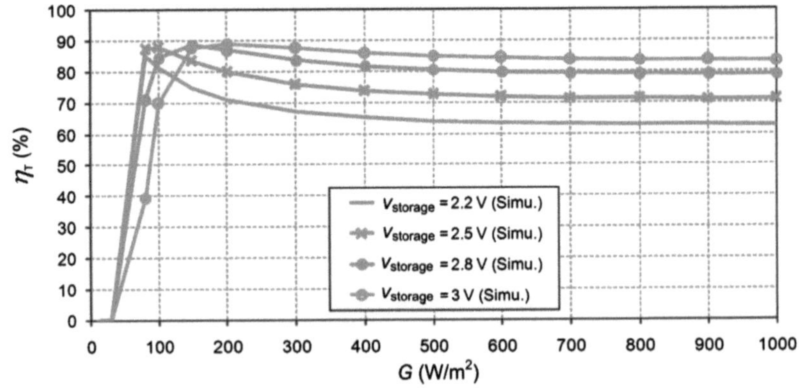

Fig. 5.15 Computed η_T at different values of $v_{storage}$ and solar irradiances. Overheating in the PV panel was not considered

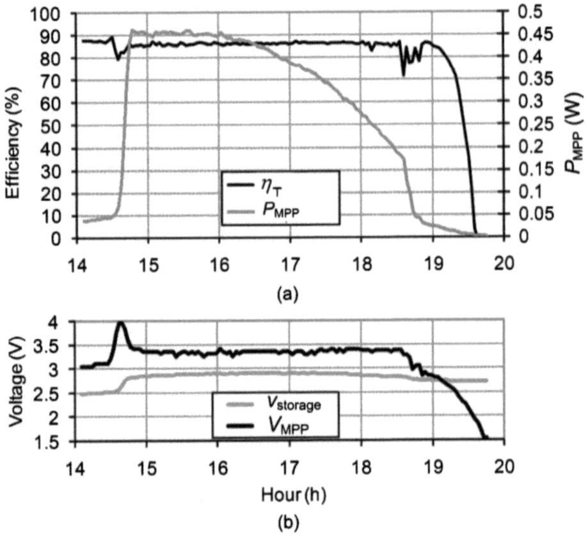

Fig. 5.16 (**a**) Measured P_{MPP} and η_T and (**b**) $v_{storage}$ and V_{MPP} changes in a direct-coupled circuit over six hours on a sunny afternoon in September

source (E3631A from Agilent) and the I–V characteristic was obtained, using a procedure similar to that described in Sect. 5.2. The PV panel was connected to the direct-coupled circuit for two and a half minutes, and then thirty seconds were left for the circuit to stabilize. To obtain the I–V characteristic, 20 I–V points were measured around V_{MPP}. To let the PV panel stabilize, an interval of ten seconds was used between measurement of two consecutive points on the I–V curve. This procedure was repeated throughout the day (from 2:00 p.m. to 8:00 p.m.). Finally, η_T was computed by dividing $P_{storage}$ by the P_{MPP} obtained from each I–V characteristic.

Figure 5.16(a) shows the experimental results for the temporal evolution of P_{MPP} and the computed η_T, which was approximately 86% at most irradiance values, but which decreased sharply at low values ($<$ *ca.* 50 W/m^2). This behavior matches the

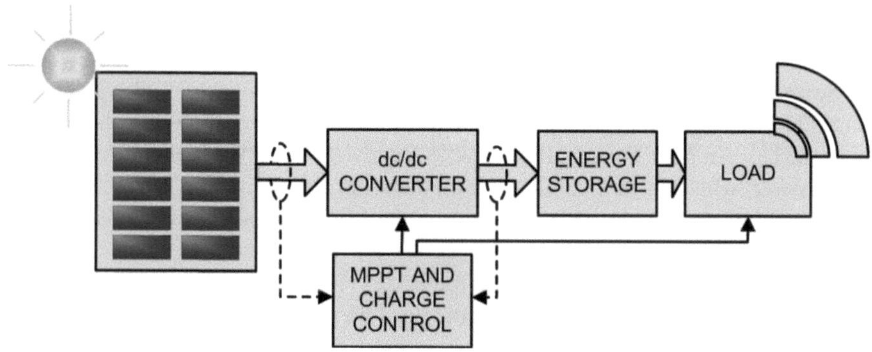

Fig. 5.17 Block diagram of an optical energy harvester equipped with an MPPT

curve of $v_{storage} = 3.0$ V in Fig. 5.14 and in Fig. 5.15. Figure 5.16(b) illustrates the temporal evolution of $v_{storage}$ and of V_{MPP}. When the experiment was started (before 2:30 p.m.), V_{MPP} was roughly 0.5 V greater than $v_{storage}$, resulting in a high η_T. In this situation, v_s was probably near V_{MPP} and η_T was mainly affected by $P_{loss,diode}$. When the sun first illuminated the PV module (at approximately 2:30 p.m.), T_{cell} was low, but the irradiance was high, resulting in a high V_{MPP}; after several minutes, the PV module was overheated by the sun (*i.e.* effect of *NOCT*) and V_{MPP} had gradually decreased. η_T fluctuated with this rapid shift in ambient conditions, underscoring how sensitive this parameter is to the environment. At the end of the day, η_T decreased once V_{MPP} had crossed $v_{storage}$.

Maximizing the energy stored in the batteries at the end of the day generally requires a thorough analysis of the η_T resulting from the expected irradiance conditions.

5.4 MPPT Circuits and Methods

Extracted power can be maximized operating the PV panel at its MPP. As the MPP varies with irradiance and temperature, an MPP tracker (MPPT) must be used. However, its implementation in autonomous sensors requires reducing its circuit consumption below the energy gain achieved by tracking the MPP. An MPPT includes a dc/dc converter and an MPPT control algorithm (Fig. 5.17). The dc/dc converter matches the output of the solar panel to the storage unit, whereas the MPPT controller forces the solar panel to work at (or near) its MPP regardless of the load, irradiance level or temperature.

The dc/dc converter can be implemented via typical power conversion circuits (see Chap. 2), such as inductor based dc/dc converters [14, 15] or circuits based on switched capacitors [16]. Inductor based dc/dc converters feature high conversion efficiencies (typically > 90%) at heavy loads, but their efficiency decreases at lighter loads. The main advantage of switched capacitors is their integration compatibility, which stems from the fact that they do not require inductors. Nonetheless, typical structures are optimized for a fixed conversion ratio. Finally, the topology employed (*i.e.* step up, step down, or step up/down) depends on the relationship between the voltages of the PV module and of the storage unit.

Several MPPT methods have been proposed and applied to high-power systems [17], and were very recently reported for use in low-power PV panels (< 1 W) [14, 15, 18]. An MPPT controller suitable for low-power PV panels must consume little power, which in turn demands computational simplicity and low-power components. Thus, to achieve a net power gain, the MPPT must address challenges not faced with high-power systems.

There are two categories of MPPT methods [19]: *direct* and *indirect*. Direct (or *true-MPPT*) methods are closed-loop and need no prior knowledge of the PV module in order to track the MPP. In contrast, indirect methods are open-loop and based on empirical databases or mathematical functions used to estimate the MPP.

One of the simplest indirect MPPT methods is *fractional open-circuit voltage (FOCV)*, which exploits the nearly linear relationship between V_{OC} and V_{MPP} of the solar cell at different levels of irradiance and at different temperatures. This result is based on observations and must be empirically determined for each type of solar cell [20]. The voltage V_{OC} is either measured periodically (by momentarily opening the output of the photovoltaic panel), as in [18], or by using a pilot cell (*i.e.* an additional solar cell of the same type in open circuit voltage configuration), as in [15] or [21]. Related methods exploit the linear relationship between I_{MPP} and I_{SC} or use a light sensor to determine irradiance conditions and then find the MPP [22].

Among direct MPPT methods, the most popular is *Perturb and Observe (P&O)*. In [14], the authors proposed a low-power implementation for this method: basically, they estimate the output power of the PV panel, and then slightly perturb the operating voltage of the solar cell to see how the power changes. If the power increases, then the perturbation should be kept in the same direction; otherwise, it should be reversed. To compute power, current and voltage usually must be measured. Compared to FOCV, P&O is advantageous because it enables a true MPPT, although at the expense of greater circuit complexity and usually, more inherent power consumption.

5.4.1 Dc/dc Converters Based on PFM Techniques

PWM-based switching converters have been used in large-scale PV modules. Nevertheless, their low efficiency at light loads make them inappropriate for low-power PV panels. Thus, we propose using pulse frequency modulation (PFM) (see Chap. 2, Sect. 2.2.4 for further details on PFM) for the dc/dc converter stage. In fact, other researchers have already developed a similar approach [14]. An input capacitor (C_{in}) was connected in parallel with the PV panel and an external hysteresis comparator was placed between the panel output and the feedback terminal (FB) of the dc/dc converter (Fig. 5.18). The shutdown terminal can also be used instead to further reduce power consumption. First, during a time T_{charge}, the switches M1 and M2 of the dc/dc converter are off, and the PV panel current (i_s) charges C_{in} until the panel

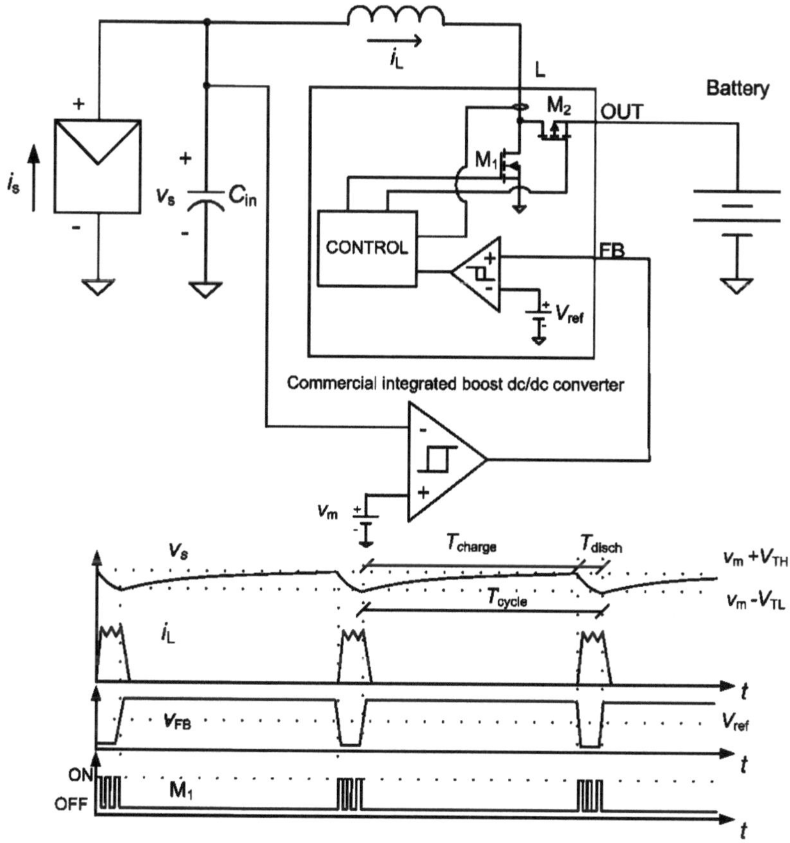

Fig. 5.18 Pulse frequency modulation applied to the dc/dc converter of the MPPT

voltage (v_s) reaches $v_m + V_{TH}$. Then, V_{FB} toggles, and M1 and M2 alternately activate, during a time T_{disch}, to transfer charge from C_{in} to the output until v_s decreases to $v_m - V_{TL}$, thereby restarting the process (each T_{cycle}). The voltage operating point of the PV panel is determined by v_m. To reach the MPP, v_m must be obtained from an MPPT controller. In Fig. 5.18 an inductor-based boost converter is shown, but the same concepts can be applied to charge pumps, and buck- or buck-boost converters.

5.4.2 Analysis

Optimal design of an MPPT system requires theoretical analysis of the static and dynamic performance of the energy conditioning stage. Hereafter, the dc/dc converter employed is assumed to be based on PFM (Sect. 5.4.1).

Static performance can be assessed with the overall power efficiency (η_T), which generally encompasses three terms:

$$\eta_T \equiv \eta_{MPPT} \cdot \eta_{av} \cdot \eta_c \tag{5.9}$$

whereby $\eta_{MPPT} \equiv \frac{P_s(v_m)}{P_{MPP}}$, $\eta_{av} \equiv \frac{\bar{P}_s}{P_s(v_m)}$, $\eta_c \equiv \frac{\bar{P}_{out}}{\bar{P}_s}$; $P_s(v_m)$ is the PV power at v_m; $\bar{P}_s$ is the average power that the panel delivers near v_m (because of the PFM applied to the dc/dc converter); and $\bar{P}_{out}$ is the average power at the output of the dc/dc converter. η_{MPPT} depends on the MPPT algorithm used; η_{av} is a general term and independent of the MPPT algorithm, and applies whenever a PFM scheme for the dc/dc converter is used; and η_c depends on the efficiency of the dc/dc converter, and on the MPPT algorithm and its particular implementation. The dynamic performance can be evaluated with the time response of the system and depends on the MPPT algorithm used and its implementation.

This section ends with an analysis of η_{av}, η_c, and general aspects of the dynamic performance. η_{MPPT} and the particularities of the dynamic performance are analyzed together with descriptions of specific MPPT methods and circuits (see Sects. 5.5 and 5.6).

5.4.2.1 Averaging Efficiency (η_{av})

The proposed implementation of the dc/dc converter using the PFM technique implies that P_s fluctuates around $P_s(v_m)$, and consequently, $\bar{P}_s < P_s(v_m)$ whenever $v_m \approx V_{MPP}$. Since the converter is inactive during T_{charge}, the output power of the PV panel (P_s) equals the incoming power at C_{in}. Thus:

$$P_s = v_s i_s = v_s C_{in} \frac{dv_s}{dt}. \tag{5.10}$$

The average power gives:

$$\bar{P}_s = \frac{1}{T}\int_{t_0}^{t_0+T} P_s\,dt = \frac{C_{in}}{T}\int_{v_s(t_0)}^{v_s(t_0+T)} v_s\,dv_s, \tag{5.11}$$

whereby t_0 and T are an arbitrary moment in time and a time interval within T_{charge}, respectively. We computed $\bar{P}_s$ during T_{charge} by using (5.11), as:

$$\bar{P}_s = \frac{1}{T}\int_{T_{charge}} P_s\,dt = \frac{C_{in}}{T}\int_{v_m-V_h}^{v_m+V_h} v_s\,dv_s = \frac{2C_{in}v_m V_h}{T_{charge}} \tag{5.12}$$

whereby we assumed that $V_h = V_{TH} = V_{TL}$. Running (5.10) under the assumption that $v_m \approx V_{MPP}$, we obtained the following expression:

$$T_{charge} = \int_T d\tau = \int_{V_{MPP}-V_h}^{V_{MPP}+V_h} \frac{v_s C_{in}}{P_s}\,dv_s. \tag{5.13}$$

Near the MPP, the P–V curve can be approximated by a second-order Taylor polynomial:

$$P_s(v_s) \approx P_{MPP} + \left.\frac{dP_s}{dv_s}\right|_{V_{MPP}}(v_s - V_{MPP}) + \frac{1}{2}\left.\frac{d^2P_s}{dv_s^2}\right|_{V_{MPP}}(v_s - V_{MPP})^2, \tag{5.14}$$

whereby $dP_s/dv_s|_{V_{MPP}} = 0$ and $d^2P_s/dv_s^2|_{V_{MPP}} < 0$ because the MPP is a maximum. We replaced P_s in (5.13) with (5.14) and solved the integral in (5.13). We then approximated the result of the integral using a third order Taylor polynomial, obtaining:

$$T_{charge} \approx \frac{2C_{in}V_{MPP}V_h}{P_{MPP}}\left(1 - \frac{1}{6}\frac{V_h^2}{P_{MPP}}\left.\frac{d^2P_s}{dv_s^2}\right|_{V_{MPP}}\right). \tag{5.15}$$

Replacing T_{charge} in (5.12) by (5.15), using this $\bar{P}_s$ in η_{av} and considering $P_s(v_m) \approx P_{MPP}$, we arrived at:

$$\eta_{av} \approx 1 + \frac{1}{6}\frac{1}{P_{MPP}}\left.\frac{d^2P_s}{dv_s^2}\right|_{V_{MPP}} V_h^2. \tag{5.16}$$

A larger value of V_h diminishes η_{av} because of the larger variations in P_s near $P_s(v_m)$. From Fig. 5.4, the term $d^2P_s/P_{MPP}\,dv_s^2|_{V_{MPP}}$ can be inferred to be nearly constant, except at the lower end of the incident irradiance, where it increases (and η_{av} decreases). Furthermore, this term decreases as the number of serial solar cells (N) in the PV panel increases. Therefore, a larger V_h can be used with higher N panels without degrading η_{av}.

5.4.2.2 Efficiency of the Circuit (η_c)

The dc/dc converter is considered to have periodic operation, as explained in Sect. 5.4.1. The average output power is given by:

$$\overline{P}_{out} = \overline{P}_s - \overline{P}_L, \tag{5.17}$$

whereby $\overline{P}_L$ accounts for the average power losses of the dc/dc converter ($\overline{P}_{L,conv}$) and the MPPT control circuit ($\overline{P}_{L,MPPT}$).

$\overline{P}_{L,conv}$ can be expressed as:

$$\overline{P}_{L,conv} = \frac{P_{L,conv\text{-}off}T_{charge} + P_{L,conv\text{-}on}T_{disch}}{T_{cycle}}, \tag{5.18}$$

whereby $P_{L,conv\text{-}on}$ and $P_{L,conv\text{-}off}$ are the power losses when the converter is active (T_{disch}) and inactive (T_{charge}), respectively; and T_{cycle} is the sum of T_{charge} and T_{disch}.

Alternatively, $\overline{P}_{L,MPPT}$ can be expressed with the generic expression [23]:

$$\overline{P}_{L,MPPT} = P_{L,MPPT\text{-}dc} + \frac{K}{T_{cycle}}, \tag{5.19}$$

whereby $P_{L,MPPT\text{-}dc}$ accounts for the static power losses and the remaining term, in which K is a constant, accounts for the increased switching activity of the MPPT controller as T_{cycle} decreases.

Substituting (5.12), (5.17), (5.18) and (5.19) into η_c gives:

$$\eta_c \approx 1 - \left[\frac{P_{L,MPPT\text{-}dc} + P_{L,conv\text{-}off}}{\overline{P}_s} + \frac{(P_{L,conv\text{-}on} - P_{L,conv\text{-}off})T_{disch}}{T_{cycle}\overline{P}_s} + \frac{KT_{charge}/T_{cycle}}{2C_{in}V_hV_{MPP}}\right]. \tag{5.20}$$

The power balance in C_{in} during T_{cycle} can be expressed as:

$$\overline{P}_{\text{disch}} T_{\text{disch}} = \overline{P}_{\text{s}} T_{\text{cycle}}, \tag{5.21}$$

whereby $\overline{P}_{\text{disch}}$ is the average power transferred from the capacitor to the converter during T_{disch}. Thus, (5.20) can be rearranged to

$$\eta_{\text{c}} \approx 1 - \left[\frac{P_{\text{L,MPPT-dc}} + P_{\text{L,conv-off}}}{\overline{P}_{\text{s}}} + \frac{(P_{\text{L,conv-on}} - P_{\text{L,conv-off}})}{\overline{P}_{\text{disch}}} + \frac{\text{K}(1 - \overline{P}_{\text{s}}/\overline{P}_{\text{disch}})}{2C_{\text{in}} V_{\text{h}} V_{\text{MPP}}} \right]. \tag{5.22}$$

For low $\overline{P}_{\text{s}}$ values, the contribution of the power losses of the MPPT control circuit ($P_{\text{L,MPPT-dc}}$) and converter ($P_{\text{L,conv-off}}$) become significant. Contrariwise, for high $\overline{P}_{\text{s}}$ values, the efficiency of the converter ($\frac{\overline{P}_{\text{L,conv-on}}}{\overline{P}_{\text{disch}}}$) becomes dominant. At any $\overline{P}_{\text{s}}$, η_{c} increases as the value of the product $C_{\text{in}} V_{\text{h}}$ increases.

5.4.2.3 Dynamic Performance

Response time is influenced by the characteristics of the PV panel, C_{in}, and the MPPT control algorithm. Regardless of the MPPT algorithm used, T_{cycle} (and therefore, T_{charge} and T_{disch}) must be considered. Analyzing the circuit presented in Fig. 5.18, and again considering $V_{\text{h}} = V_{\text{TH}} = V_{\text{TL}}$, gives:

$$T_{\text{charge}} = \frac{2C_{\text{in}} V_{\text{h}}}{\overline{i_{\text{s}}}}, \tag{5.23}$$

$$T_{\text{disch}} = \frac{2C_{\text{in}} V_{\text{h}}}{(\overline{i_d} - \overline{i_{\text{s}}})}, \tag{5.24}$$

whereby $\overline{i_{\text{s}}}$ is the average value of the PV panel current during T_{cycle} and $\overline{i_d}$ is the average input current demanded by the dc/dc converter during T_{disch}. For the sake of simplicity, we considered the same $\overline{i_{\text{s}}}$ for both T_{charge} and T_{disch}. The value of $\overline{i_{\text{s}}}$ must be higher than zero in order to have a finite value of T_{charge}, and lower than $\overline{i_d}$ in order to have a positive and finite value of T_{disch}.

Combining (5.23) and (5.24) gives:

$$T_{\text{cycle}} = T_{\text{charge}} + T_{\text{disch}} = 2C_{\text{in}} V_{\text{h}} \frac{\overline{i_d}}{\overline{i_{\text{s}}}(\overline{i_d} - \overline{i_{\text{s}}})}. \tag{5.25}$$

The maximum value of T_{cycle} (T_{max}) must be limited to achieve a limited time response, whereas the minimum values of T_{charge} and T_{disch} (T_{min}) must be limited to a reasonable value (*i.e.* T_{min} should be sufficiently higher than the switching times of the comparator and dc/dc converter).

Whenever $\overline{i_{\text{s}}}(\text{min}) < \overline{i_d} - \overline{i_{\text{s}}}(\text{max})$ (whereby $\overline{i_{\text{s}}}(\text{min})$ and $\overline{i_{\text{s}}}(\text{max})$ are the minimum and maximum values considered for $\overline{i_{\text{s}}}$, respectively) and $\overline{i_{\text{s}}}(\text{min}) \ll \overline{i_d}$, T_{cycle}

will be maximal at $\overline{i_s}(\min)$ and roughly equal to T_{charge}. Thus, using (5.23), the following expression is obtained for the maximum value of C_{in}:

$$C_{\text{in,max}} = \frac{T_{\max} \cdot \overline{i_s}(\min)}{2V_{\text{h}}}. \tag{5.26}$$

When the irradiance is null, $\overline{i_s} = 0$. In practice, this means that whenever the PV panel generates $\overline{i_s} < \overline{i_s}(\min)$, T_{cycle} will be longer than $T_{\max}$. Or, equivalently, (5.26) states that an arbitrarily low value of $\overline{i_s}(\min)$ cannot be chosen in order to limit the selected C_{in} to a feasible value. For $\overline{i_s}(\min)$, there is also a minimum value of T_{disch}, which is lower than the minimum value of T_{charge}, which in turn occurs at $\overline{i_s}(\max)$. Thus, from (5.24), the following expression is obtained for the minimum value of C_{in}:

$$C_{\text{in,min}} = \frac{T_{\min} \cdot \overline{i_d}}{2V_{\text{h}}}. \tag{5.27}$$

Thus, in order for $C_{\text{in,max}}$ to be greater than $C_{\text{in,min}}$, the following condition must be met:

$$\frac{\overline{i_d}}{\overline{i_s}(\min)} < \frac{T_{\max}}{T_{\min}}. \tag{5.28}$$

In practice, $\overline{i_d}$ and $T_{\min}$ (and therefore, (5.27) and (5.28)) may not be clearly solved. Nonetheless, replacing $\overline{i_d}$ in (5.28) with $\overline{i_s}(\max)$, leads to the conclusion that the considered dynamic range for $\overline{i_s}$ cannot be arbitrarily large and must be bounded.

Whenever $\overline{i_s}(\min) > \overline{i_d} - \overline{i_s}(\max)$ and $\overline{i_s}(\min) \ll \overline{i_d}$ holds, $\overline{i_s}(\min)$ in (5.26) and (5.28) must be substituted with $\overline{i_d} - \overline{i_s}(\max)$; however, (5.27) remains unchanged.

When considering a particular MPPT method, other constraints can apply to (5.26), (5.27) and (5.28).

5.5 LOCV Method

This section presents a simple and high-efficient open-loop MPPT method for PV panels. The method, referred herein as Linear Open Circuit Voltage (LOCV), is based on estimating the maximum power point from a linear fit of the open circuit voltage [24]. The generic model for PV cells is used in order to compute the tracking efficiency versus irradiance for several temperatures. Then, experimental tests are performed using a 500 mW PV panel. Tracking efficiency of the LOCV method is compared against the popular Fractional Open Circuit Voltage (FOCV) method. The proposed LOCV method outperforms the FOCV method and achieves efficiencies comparable to that of the best and more complex MPPT methods. Finally, a particular implementation is proposed for both FOCV and LOCV methods.

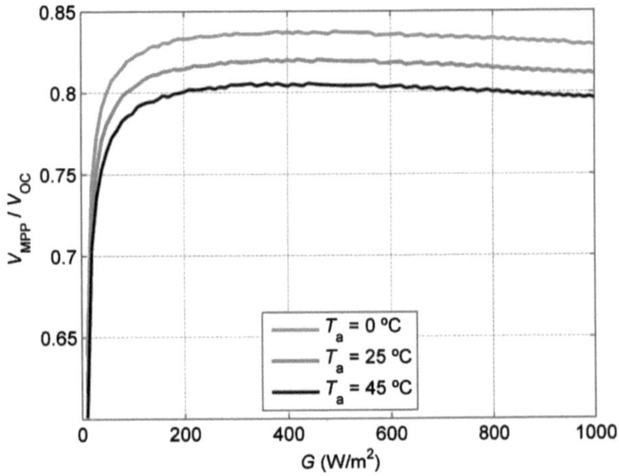

Fig. 5.19 $V_{\mathrm{MPP}}/V_{\mathrm{OC}}$ plotted against G at three values of T_{a}

5.5.1 *Efficiency*

FOCV methods are based on the ratio $V_{\mathrm{MPP}}/V_{\mathrm{OC}}$, which depends on the fabrication technology and materials used for the solar cell and on the meteorological conditions [19]. Figure 5.19 shows computed values of this ratio plotted against G at three values of T_{a} for the PV model described in Sect. 5.1. The parameter values used for these simulations are the same as those in Fig. 5.2. As observed in the figure, the ratio remains nearly constant at high irradiance levels and at a given T_{a}, but drops at low irradiances values and with increasing T_{a}.

All the FOCV methods found in the literature propose calculating V_{MPP} as:

$$V_{\mathrm{MPP}} \equiv k V_{\mathrm{OC}} \tag{5.29}$$

whereby k is a constant whose value should be set following thorough characterization of the PV panel under varying meteorological conditions (although most authors only consider irradiance variations [18, 25]). Obviously, from the results of Fig. 5.19, the PV panel will only work at its MPP in a limited range of irradiance levels and temperatures.

Figure 5.20(a) shows computed values of V_{MPP} plotted against V_{OC} at several values of T_{a} for a single solar cell. The graph also includes a linear regression line given by:

$$V_{\mathrm{MPP}} = a V_{\mathrm{OC}} + b \tag{5.30}$$

whereby $a = 0.906$ and $b = -48$ mV. The correlation coefficient (R) is equal to 0.997, indicating that the computed values fit well with the regression line ($R = 1$ indicates that the regression line perfectly fits the data). Figure 5.20(b) shows the ratio $V_{\mathrm{MPP}}/V_{\mathrm{OC}}$ for a PV module with $N = 8$. As observed in the figure, the value of a is the same as in the case of Fig. 5.20(a), whereas the value of b is eight times that for one single cell. For a PV panel with N cells, the value of b is generally N times that for one single cell. To the best of our knowledge, (5.30) has never

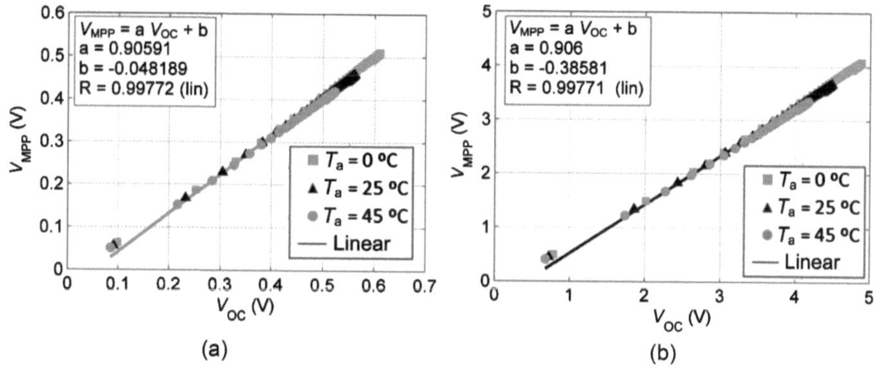

Fig. 5.20 V_{MPP} plotted against V_{OC}, with the corresponding linear regression for (**a**) one cell and (**b**) eight cells connected in series

before been proposed for computing V_{MPP} from V_{OC}. The proposed LOCV method is based on the use of (5.30).

We assessed and compared the efficiency of the FOCV and LOCV MPPT algorithm (η_{MPPT}) in function of G for a single solar cell. Figure 5.21(a) and (b) show the computed η_{MPPT} using the FOCV method at three different values of k (0.77, 0.80 and 0.82) and $T_a = 25$ °C, and at three different temperatures (0 °C, 25 °C and 45 °C) with $k = 0.82$, respectively. A k of 0.82 offers the highest η_{MPPT} at medium to high irradiances, although at the cost of lower η_{MPPT} (*ca.* 95%) at low irradiances (10 W/m^2). Alternatively, $k = 0.77$ presents higher η_{MPPT} at low irradiances (near 100% at $G = 30$ W/m^2), but drops to 98% for medium irradiances. Finally, η_{MPPT} varies by 0.5% for different values of T_a, from 0 °C to 45 °C. Figure 5.22 shows the computed η_{MPPT} using the LOCV method at several values of T_a. In this case, η_{MPPT} is greater than 99.8% for almost all the input range of irradiances and all the computed temperatures: indeed, only for very low irradiances ($G = 10$ W/m^2), does it drop to 99.5%. Thus, the LOCV method outperforms the FOCV method. Furthermore, the FOCV method requires a careful selection of the parameter k for improved efficiency.

5.5.2 Experimental Characterization

Figure 5.23 shows the efficiency of the FOCV method obtained with the PV array simulator described in Sect. 5.2. The results are consistent with those presented in Fig. 5.21(a). A lower minimum efficiency was achieved at $k = 0.8$ and $k = 0.82$ with lower irradiances, although the equivalent G is higher than that shown in Fig. 5.21(a).

The linear regression in Fig. 5.20 does not adequately fit the data of the PV array simulator because it has a different *NOCT* and V_{OCr} than those used for the

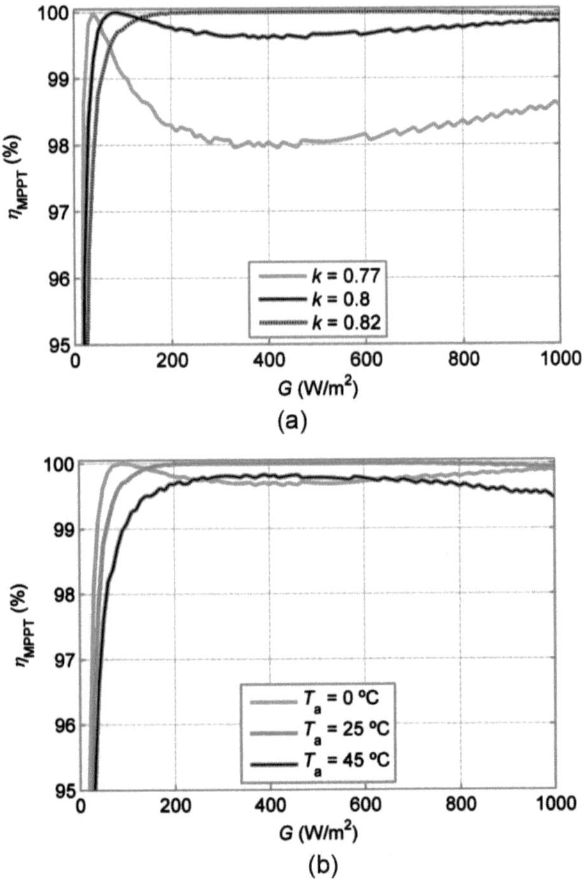

Fig. 5.21 η_{MPPT} calculated using the FOCV method and plotted against G for a single solar cell at (**a**) three values of k and $T_a = 25\,°C$ and (**b**) at three values of T_a and $k = 0.82$

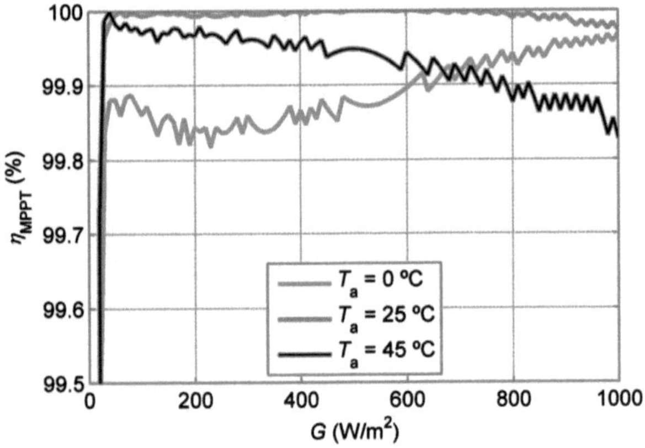

Fig. 5.22 η_{MPPT} calculated using the LOCV method and plotted against G at three values of T_a

simulations. Thus, the data of the PV array simulator (Fig. 5.24(a)) were refit, and the resulting η_{MPPT} was very good ($> 99.9\%$) throughout the range of P_{MPP}

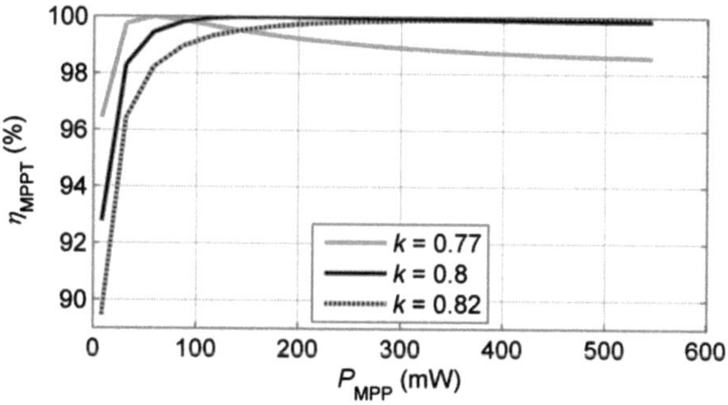

Fig. 5.23 η_{MPPT} calculated using the FOCV method and plotted against P_{MPP} at three different values of k for the PV array simulator described in Sect. 5.2

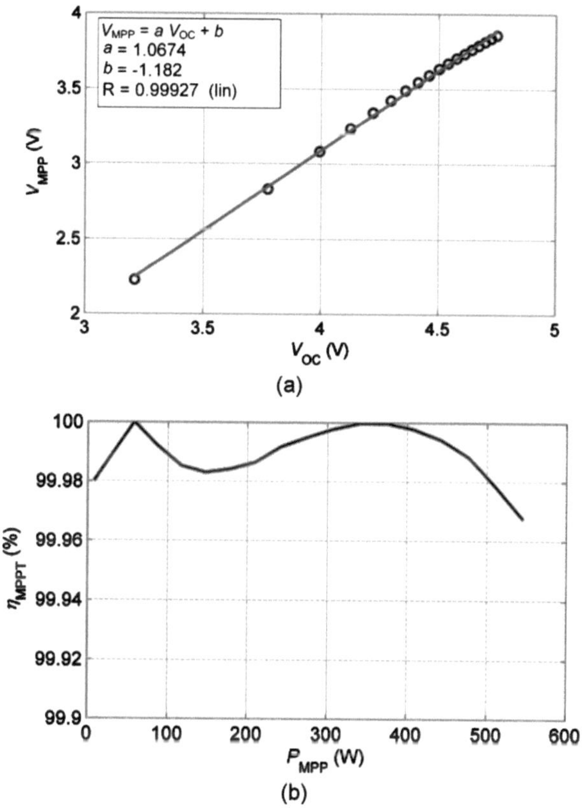

Fig. 5.24 Simulation results for the PV array simulator described in Sect. 5.2: (**a**) linear fit of V_{MPP} plotted against V_{OC}, and (**b**) η_{MPPT} calculated using the LOCV method and plotted against P_{MPP}

(Fig. 5.24(a)). Nonetheless, this fit considered only one value of T_a; thus, broader characterization must be performed in order to find the optimal fit for several values of T_a.

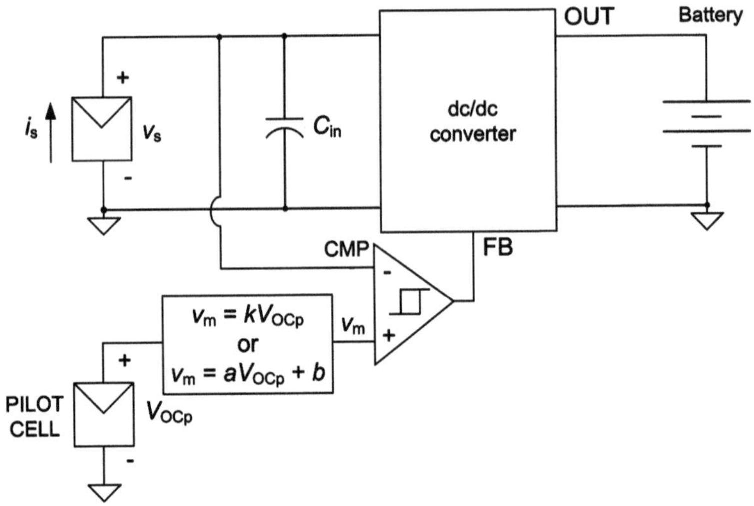

Fig. 5.25 Block diagram of the proposed implementation

5.5.3 Implementation and Time Response

The FOCV and the LOCV methods can be implemented by opening the output of the PV panel periodically and measuring the V_{OC}, which causes some energy loss [18]. Alternatively, an additional (pilot) solar cell of the same type can be used to determine the V_{MPP} of the target PV panel from its V_{OC} (V_{OCp}) [21]. Here, the overall cost increases, and any difference in the characteristics, irradiation level or temperature between the two cells will result in errors during calculation of V_{MPP}. An implementation with a pilot cell is presented in this section.

Figure 5.25 shows the block diagram of a proposed implementation using a pilot cell. The value of v_m is obtained from V_{OCp} by using a stage that implements either (5.29) or (5.30). This stage must present a high-impedance input to achieve open-circuit conditions for the pilot cell. Thus, as explained in Sect. 5.4.1, v_s oscillates near v_m, which ideally is equal or at least close to V_{MPP}. Whenever the pilot cell contains a different number of serial cells than does the main PV panel, either (5.29) or (5.30) must be accordingly scaled. Static performance can be assessed based on the overall power efficiency, as described in Sect. 5.4.2. η_{av} and η_c can be calculated as in Sects. 5.4.2.1 and 5.4.2.2, respectively, and η_{MPPT}, will depend on the particular MPPT method, either FOCV or LOCV.

Figure 5.26 shows the expected waveform of v_s and v_m for the case in which V_{MPP} first increases and then decreases due to a change of irradiance or temperature. As observed in the figure, the converter oscillates near v_m, and when V_{OCp} changes, a new v_m is obtained, whose value C_{in} is charged or discharged to.

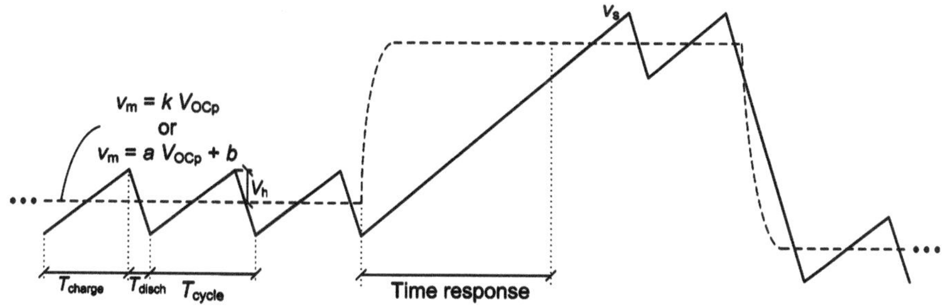

Fig. 5.26 Expected v_s and v_m waveforms when V_{MPP} changes

Apart from the considerations of Sect. 5.4.2.3, the time response of the system due to an increase of V_{MPP} (t_{pilot_cell}) can be approximated by:

$$t_{pilot\_cell} = \frac{C_{in}(V_{MPP2} - V_{MPP1})}{I_{MPP2}}, \tag{5.31}$$

whereby V_{MPP1} and V_{MPP2} correspond to the V_{MPP} values before and after the V_{MPP} change, respectively, and I_{MPP2} corresponds to V_{MPP2}. A similar approach can be taken when V_{MPP} decreases, giving:

$$t_{pilot\_cell} = \frac{C_{in}(V_{MPP1} - V_{MPP2})}{\overline{i_d} - I_{MPP2}}. \tag{5.32}$$

Considering the plot of V_{MPP} against P_{MPP} shown in Fig. 5.2, the worst dynamic response will be found at low currents of the PV panel (*i.e.* at low irradiations and/or with increasing V_{MPP}).

5.5.4 Assigning Parameter Values

To obtain a high η_T at a given dynamic response, the values of several parameters must be appropriately selected. Firstly, in order to achieve a high η_{MPPT}, a proper value of k (for the FOCV method), or of a and of b (for the LOCV method), must be selected. Equation (5.16) generates a high boundary for V_h. Thus, by limiting t_{pilot_cell} and T_{charge}, a high limit for C_{in} is obtained. Conversely, the product $C_{in} V_h$ must be high enough to maximize η_c as well as to limit the minimum values of T_{charge} and of T_{disch} (see Sect. 5.4.2.3).

For the PV array simulator described in Sect. 5.2, we chose parameter values by considering a given range of the PV panel power and the corresponding I_{MPP}. From Sect. 5.2, $I_{MPP,min} = 3.7$ mA and $I_{MPP,max} = 141.4$ mA. By fixing $t_{pilot\_cell} < 1$ s and solving (5.31) at a low $\overline{P_s}$ with $V_{MPP1} = 2.226$ V, $V_{MPP2} = 2.832$ V and $I_{MPP2} = 11.1$ mA results in $C_{in} < 18.3$ mF. Figure 5.27 illustrates (5.16). For the calculus, PV power-related parameters (*e.g.* the second derivative of P_s, and P_{MPP}) were estimated from the experimental P–V curve of the PV array simulator. At

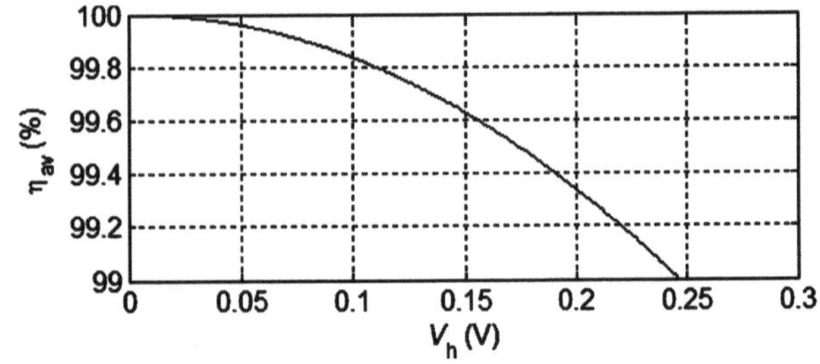

Fig. 5.27 Computed η_{av} versus V_h

$V_h = 100$ mV, $\eta_{av} > 99.8\%$. A smaller value of V_h is not recommended, in order to minimize the effect of the trigger noise at the input of the hysteresis comparator (Fig. 5.25). By selecting $C_{in} = 1$ mF, $t_{pilot\_cell} = 54$ ms. T_{charge}, calculated from (5.23), will vary between 1.4 ms at $I_{MPP,max}$ and 54 ms at $I_{MPP,min}$. Higher values of C_{in} will lead to longer times and larger capacitors. Capacitors with low leakage and series resistance are preferred.

5.6 A Novel Closed-Loop MPPT Technique

Classical[1] closed-loop MPPT techniques require the measurement of current and voltage of the PV panel in order to estimate the power, which is used as an input parameter of the MPPT controller. This section introduces a new technique, previously reported in [26], and [27] suitable for low-power (< *ca.* 1 W) PV panels that avoids the measurement of current. Further, an ad hoc MPPT algorithm is also implemented. The static and dynamic performance is first theoretically analyzed and design criteria are provided. A prototype was implemented with a 500 mW PV panel, a commercial boost converter, and low-power components for the MPPT controller. Laboratory measurements were performed to assess the effectiveness of the proposed technique. Tracking efficiency was higher than 99.6%. Overall efficiency was higher than 92% for a PV panel power higher than 100 mW. This is in part feasible due to the low power consumption of the MPPT controller, which was kept lower than 350 μW. The time response of the tracking circuit was tested to be around 1 s. Field measurements showed energy gains higher than 10.3% with respect to a direct-coupled solution for an ambient temperature of 26 °C. Higher gains are expected for lower temperatures.

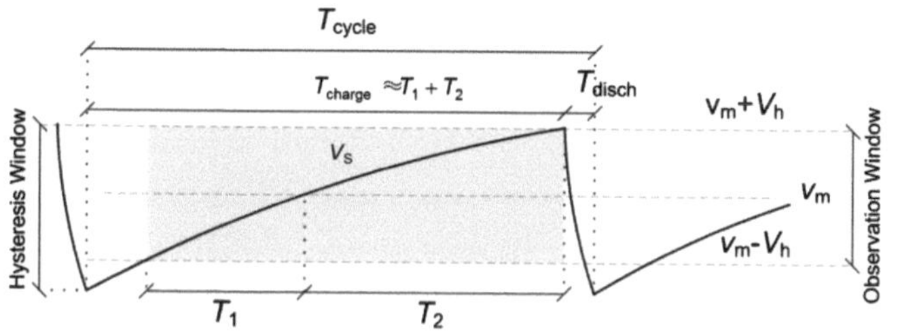

Fig. 5.28 Hysteresis window and observation window of the proposed MPPT method

5.6.1 *Theoretical Approach*

A closed-loop MPPT method must measure the generated power of the PV panel, and then find and follow its maximal value. Herein we propose using the circuit described in Fig. 5.18 together with a new MPPT technique that we developed that is well suited for low-power PV panels. To achieve the MPP, the method tends to null the difference of the panel's average power in two consecutive time intervals (T_1 and T_2) within T_{charge}. As Fig. 5.28 shows, T_1 and T_2 correspond to the lower and upper halves, respectively, of an observation voltage window (whose gap is $2V_{\mathrm{h}}$) centered on the operating voltage v_{m}. As observed in the figure, the minimum voltage of the observation window is slightly higher than the minimum voltage of the hysteresis window, enabling the devices used in the MPPT controller (see Sect. 5.6.2) to overcome the transient response produced by the fast discharge of C_{in}. However, the upper limits of these two windows match. For the sake of simplicity, in the rest of this section, both windows are assumed to match and T_{charge} is assumed to be roughly equivalent to $T_1 + T_2$.

Referring to Fig. 5.28, the period of v_{s} is given by:

$$T_{\mathrm{cycle}} = T_{\mathrm{charge}} + T_{\mathrm{disch}}. \tag{5.33}$$

Applying (5.11) to the calculation of the average power during T_1 and T_2 gives:

$$\begin{aligned}\overline{P}_{\mathrm{s}1} &= \frac{C_{\mathrm{in}}}{T_1}\int_{v_{\mathrm{m}}-V_{\mathrm{h}}}^{v_{\mathrm{m}}} v_{\mathrm{s}}\,dv_{\mathrm{s}} = \frac{C_{\mathrm{in}}V_{\mathrm{h}}}{T_1}\left(v_{\mathrm{m}} - \frac{V_{\mathrm{h}}}{2}\right),\\ \overline{P}_{\mathrm{s}2} &= \frac{C_{\mathrm{in}}}{T_2}\int_{v_{\mathrm{m}}}^{v_{\mathrm{m}}+V_{\mathrm{h}}} v_{\mathrm{s}}\,dv_{\mathrm{s}} = \frac{C_{\mathrm{in}}V_{\mathrm{h}}}{T_2}\left(v_{\mathrm{m}} + \frac{V_{\mathrm{h}}}{2}\right).\end{aligned} \tag{5.34}$$

From (5.34), the power difference is:

$$\Delta P_{\mathrm{s}} \equiv \overline{P}_{\mathrm{s}2} - \overline{P}_{\mathrm{s}1} = \frac{C_{\mathrm{in}}V_{\mathrm{h}}}{T_2T_1}\left(v_{\mathrm{m}}(T_1 - T_2) + \frac{V_{\mathrm{h}}}{2}(T_1 + T_2)\right). \tag{5.35}$$

The sign of (5.35) indicates the position of the operating voltage v_{m} relative to V_{MPP}. Figure 5.29 illustrates this concept, showing the characteristic power profile plotted against the operating voltage of a generic PV panel (P–V curve). As observed in the figure, whenever (5.35) is positive, v_{m} is less than V_{MPP} and should be increased. Contrariwise, whenever (5.35) is negative, v_{m} is greater than V_{MPP}

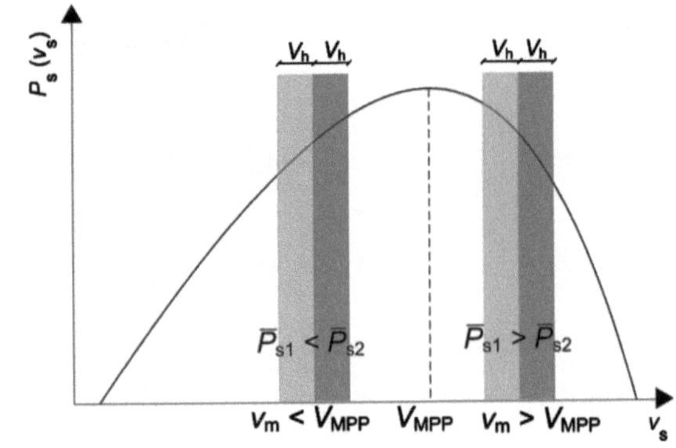

Fig. 5.29 Evaluation of the operating point (v_m) relative to the MPP (V_{MPP})

and should be decreased. The sign of (5.35) depends exclusively on the term within the parenthesis. This term is defined as $L(v_m)$ and calculated as:

$$L(v_m) \equiv v_m(T_1 - T_2) + \frac{V_h}{2}(T_1 + T_2). \tag{5.36}$$

An equilibrium point, $v_{m,eq}$, is achieved when $\Delta P_s = 0$, as shown by:

$$L(v_{m,eq}) = v_{m,eq}(T_1 - T_2) + \frac{V_h}{2}(T_1 + T_2) = 0, \tag{5.37}$$

resulting in $v_{m,eq} = V_{MPP}$ if $V_h \ll V_{MPP}$.

5.6.2 Implementation

To implement the MPPT control law, we calculated v_m as the iterative sum of $L(v_m)$ in each new cycle:

$$v_m(n) = \frac{1}{\tau_{INT}} \sum_{i=1}^{n-1} \left(\int_{T_1(i)} v_m(i)\, dt - \int_{T_2(i)} v_m(i)\, dt + \int_{T_1(i)+T_2(i)} \frac{V_h}{2} \cdot dt \right), \tag{5.38}$$

whereby $v_m(n)$ is sampled at the end of each T_{charge}; τ_{INT} is an integral constant; and n is the cycle number. As observed, (5.38) was obtained by replacing the multiplication operations in (5.36) with integral operators. This equivalence is based on the fact that V_h is a constant and the sampled v_m has a constant value during T_1 and T_2. Moreover, unlike in other methods, measurement of the current is not required. The value of v_m will increase whenever $v_m(n) < V_{MPP}$, and will decrease whenever $v_m(n) > V_{MPP}$, thus $v_{m,eq}$ will tend to be roughly equal to V_{MPP}.

A schematic of the implemented MPPT circuit is shown in Fig. 5.30. The comparator CMP_2 toggles SW_2 at the instant that v_s crosses v_m. The hysteresis comparator CMP_3 activates the converter during T_{disch}, and together with CMP_1, enables SW_2 and selects terminal 1 of SW_1 during T_1 and T_2. The output signals of SW_1 and SW_2 are added and integrated, and the resulting output (v_{INT}) is sampled

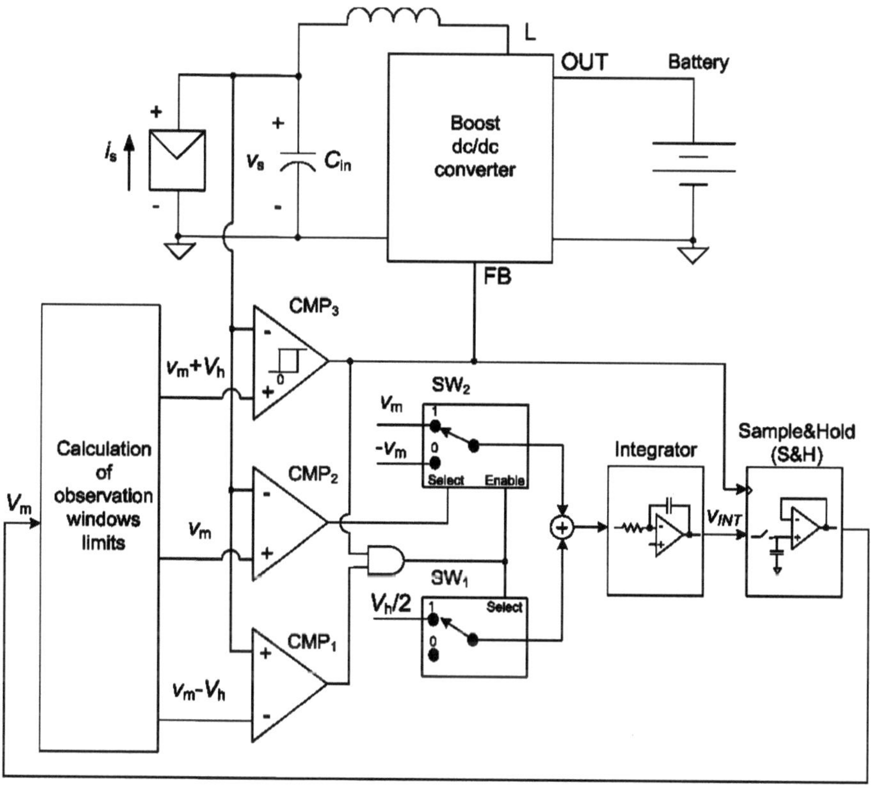

Fig. 5.30 Simplified schematic of the proposed MPPT

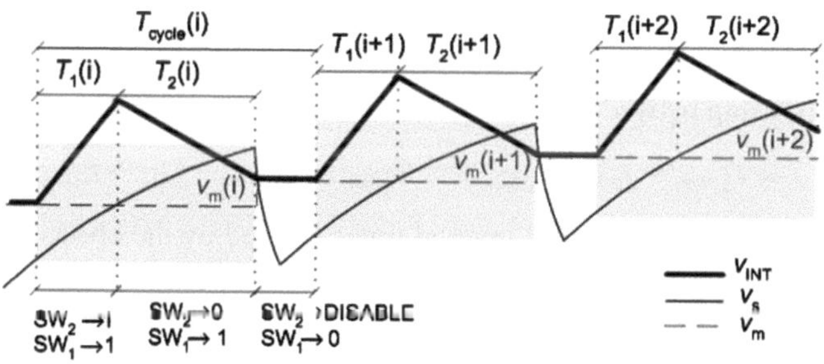

Fig. 5.31 Expected temporal evolution of v_m, v_s, and v_{INT} for the case in which $v_m < V_{MPP}$

at the end of T_{charge}, providing a new value of v_m. Figure 5.31 illustrates the expected temporal evolution of v_m, v_s, and v_{INT}, for the case in which $v_m < V_{MPP}$. As observed in the figure, v_m is updated at the end of T_{charge} and increases towards V_{MPP}. The state of SW_1 and SW_2 during a full cycle is also shown.

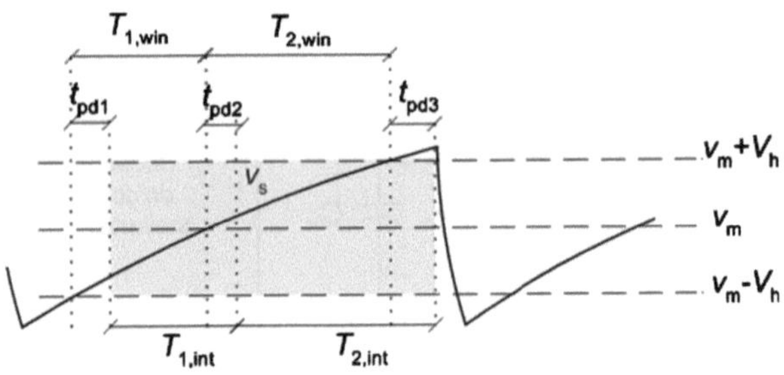

Fig. 5.32 Definition of the time intervals when the propagation delays of the comparators are considered

5.6.3 *Analysis*

To optimize design of the MPPT circuit, we theoretically analyzed the static and dynamic performance of the MPPT. Static performance was assessed based on the overall power efficiency, as described in Sect. 5.4.2 and in (5.9). η_{av} and η_c were calculated as in Sects. 5.4.2.1 and 5.4.2.2, respectively. η_{MPPT} had to be computed for the MPPT method and circuit. In contrast, the dynamic performance was assessed in function of the time constant of v_m and modeled as a first-order time function.

5.6.3.1 Efficiency of the MPPT Algorithm

We devised the MPPT controller to achieve high tracking efficiency with low power consumption. Propagation delays of the comparators CMP1 to CMP3 (Fig. 5.30) affect the tracking efficiency. This is particularly relevant, due to the design tradeoff between the power consumption and propagation delay of COTS comparators. The following sections describe analysis of how these delays affect tracking efficiency.

Figure 5.32 shows the observation window considering the propagation delays t_{pd1}, t_{pd2} and t_{pd3} introduced by CMP$_1$, CMP$_2$ and CMP$_3$, respectively.

The relationship between the different time intervals is given by:

$$T_{1,int} = T_{1,win} - t_{pd1} + t_{pd2}, \qquad T_{2,int} = T_{2,win} - t_{pd2} + t_{pd3}, \tag{5.39}$$

$T_{1,win}$ and $T_{2,win}$ are the ideal interval times defined by the observation window and provide the real ΔP_s through (5.35). $T_{1,int}$ and $T_{2,int}$ are the integration times of the MPPT circuit of Fig. 5.30 and determine $v_{m,eq}$ through (5.37). Thus, due to the propagation delays, $\Delta P_s(v_{m,eq})$ will not be zero and, consequently, $v_{m,eq}$ will not reach the MPP at the equilibrium point, thereby worsening the tracking efficiency. Regardless, the equilibrium point is still assumed to be near the MPP.

Substituting T_1 and T_2 in (5.37) with $T_{1,int}$ and $T_{2,int}$, respectively, and using (5.39), gives:

$$\begin{aligned} &v_{m,eq}\big((T_{1,win} - t_{pd1} + t_{pd2}) - (T_{2,win} - t_{pd2} + t_{pd3})\big) \\ &\quad + \frac{V_h}{2}\big((T_{1,win} - t_{pd1} + t_{pd2}) + (T_{2,win} - t_{pd2} + t_{pd3})\big) = 0. \end{aligned} \tag{5.40}$$

Thus, substituting T_1 and T_2 in (5.35) with $T_{1,\mathrm{win}}$ and $T_{2,\mathrm{win}}$, respectively, using (5.40) and considering $V_\mathrm{h} \ll v_{\mathrm{m,eq}}$, yields:

$$\Delta P_\mathrm{s}(v_{\mathrm{m,eq}}) \approx \frac{C_\mathrm{in} V_\mathrm{h}}{T_{1,\mathrm{win}} T_{2,\mathrm{win}}} v_{\mathrm{m,eq}} (t_{\mathrm{pd1}} - 2t_{\mathrm{pd2}} + t_{\mathrm{pd3}}). \tag{5.41}$$

The slope of the P–V curve at the equilibrium point can be related to the propagation delays via:

$$\left.\frac{dP_\mathrm{s}}{dv_\mathrm{s}}\right|_{v_{\mathrm{m,eq}}} \approx \left.\frac{\Delta P_\mathrm{s}}{V_\mathrm{h}}\right|_{\substack{v_{\mathrm{m,eq}} \\ V_\mathrm{h} \ll V_{\mathrm{MPP}}}} \approx \frac{t_{\mathrm{pd1}} - 2t_{\mathrm{pd2}} + t_{\mathrm{pd3}}}{C_\mathrm{in} V_\mathrm{h}^2} P_{\mathrm{MPP}} I_{\mathrm{MPP}}, \tag{5.42}$$

whereby $v_{\mathrm{m,eq}}$ was considered to be roughly equal to V_{MPP} and:

$$T_{1,\mathrm{win}} \approx T_{2,\mathrm{win}} \approx \frac{T_{\mathrm{charge}}}{2} \approx \frac{C_\mathrm{in} V_\mathrm{h}}{I_{\mathrm{MPP}}}. \tag{5.43}$$

The value of P_s near V_{MPP} can be approximated using a second-order Taylor polynomial (see (5.14)). Taking the derivative of (5.14) gives:

$$v_\mathrm{s} - V_{\mathrm{MPP}} = \frac{dP_\mathrm{s}/dv_\mathrm{s}|_{v_\mathrm{S}}}{d^2P_\mathrm{s}/dv_\mathrm{s}^2|_{V_{\mathrm{MPP}}}}. \tag{5.44}$$

Therefore, substituting (5.14) at $v_\mathrm{s} = v_{\mathrm{m,eq}}$ in η_{MPPT}, and using (5.42) and (5.44), affords:

$$\eta_{\mathrm{MPPT}} \approx 1 + \left(\frac{t_{\mathrm{pd1}} - 2t_{\mathrm{pd2}} + t_{\mathrm{pd3}}}{C_\mathrm{in} V_\mathrm{h}} \frac{I_{\mathrm{MPP}}}{V_\mathrm{h}}\right)^2 \Bigg/ \left(\frac{2}{P_{\mathrm{MPP}}} \left.\frac{d^2P_\mathrm{s}}{dv_\mathrm{s}^2}\right|_{V_{\mathrm{MPP}}}\right). \tag{5.45}$$

Notice that the second derivate of P_s is negative, and therefore, $\eta_{\mathrm{MPPT}} < 1$. The tracking efficiency will tend to unity at low and matched propagation delays, and at higher values of V_h. A higher value of the product $C_\mathrm{in} V_\mathrm{h}$ also increases the tracking efficiency. This seems obvious from (5.43), since T_{charge} increases, and consequently, the relative error contributed by the propagation delays diminishes.

Adapting this method to high-power applications would require either selecting higher values of $C_\mathrm{in} V_\mathrm{h}$ (to maintain a high tracking accuracy) or using faster comparators, although this would obviously increase the power consumption of the MPPT controller.

5.6.3.2 Dynamic Performance

The dynamic response of v_m can be described by a time-continuous function. Therefore:

$$\frac{dv_\mathrm{m}}{dt} \approx \frac{v_\mathrm{m}(n+1) - v_\mathrm{m}(n)}{T_{\mathrm{cycle}}}. \tag{5.46}$$

Using (5.35) and (5.38) gives:

$$v_\mathrm{m}(n+1) - v_\mathrm{m}(n) = \frac{T_2(n) \cdot T_1(n)}{\tau_{\mathrm{INT}} C_\mathrm{in}} \frac{\Delta P_\mathrm{s}(v_\mathrm{m}(n))}{V_\mathrm{h}}, \tag{5.47}$$

which, by using (5.12) with $T = T_1 + T_2$ and (5.43), can be rewritten as:

$$v_{\mathrm{m}}(n+1) - v_{\mathrm{m}}(n) = \frac{T_{\mathrm{charge}}(n)}{2\tau_{\mathrm{INT}}} v_{\mathrm{m}}(n) \frac{\Delta P_{\mathrm{s}}(v_{\mathrm{m}}(n))}{\overline{P}_{\mathrm{s}}(v_{\mathrm{m}}(n))}$$
$$\approx \frac{T_{\mathrm{charge}}}{2\tau_{\mathrm{INT}}} V_{\mathrm{h}} v_{\mathrm{m}} \left(\frac{1}{P_{\mathrm{s}}} \frac{dP_{\mathrm{s}}}{dv_{\mathrm{s}}} \right) \bigg|_{v_{\mathrm{m}}}. \tag{5.48}$$

Substituting (5.48) in (5.46) affords:

$$\frac{dv_{\mathrm{m}}}{dt} \approx \frac{V_{\mathrm{h}}}{2\tau_{\mathrm{INT}}(1 + T_{\mathrm{disch}}/T_{\mathrm{charge}})} v_{\mathrm{m}} \left(\frac{1}{P_{\mathrm{s}}} \frac{dP_{\mathrm{s}}}{dv_{\mathrm{s}}} \right) \bigg|_{v_{\mathrm{m}}}. \tag{5.49}$$

The lowest slope of P_{s} will be close to the MPP; therefore, the interval of time required to reach the MPP can be assumed to be chiefly due to the dynamic response in this region. Close to the MPP, this dynamic can be approximated by a first-order Taylor polynomial near V_{MPP}:

$$\frac{dv_{\mathrm{m}}}{dt} \approx \frac{1}{\tau} \cdot (V_{\mathrm{MPP}} - v_{\mathrm{m}}). \tag{5.50}$$

If $T_{\mathrm{charge}} \gg T_{\mathrm{disch}}$, then:

$$\tau \approx -2\tau_{\mathrm{INT}} \Bigg/ \left(V_{\mathrm{h}} V_{\mathrm{MPP}} \frac{1}{P_{\mathrm{MPP}}} \frac{d^2 P_{\mathrm{s}}}{dv_{\mathrm{s}}^2} \bigg|_{V_{\mathrm{MPP}}} \right). \tag{5.51}$$

Note that T_{cycle} does not appear in (5.51), and therefore, will not determine the dynamic of v_{m}. Nevertheless, T_{cycle} determines the time discretization of v_{m} (see Fig. 5.31). Thus, in order for T_{cycle} to be kept short enough, T_{charge} must be much shorter than τ. However, if T_{cycle} is not sufficiently short the reader is referred to Sect. 5.4.2.3 for the computation of the dynamic response.

5.6.4 Assigning the Parameter Values

This section describes the methodology used to assign values for C_{in}, V_{h}, and τ_{INT} that would enable maximum efficiency (η_{T}) with a given dynamic response. A value of 1 s was chosen for τ.

For the calculus, PV power-related parameters (*e.g.* the second derivative of P_{s}, and P_{MPP}) were estimated from the experimental P–V curve of the PV described in Sect. 5.2.

The proposed MPPT method was tested for a given range of the PV panel power and the corresponding I_{MPP}. From Sect. 5.2, $I_{\mathrm{MPP,min}} = 3.7$ mA and $I_{\mathrm{MPP,max}} =$ 141.4 mA.

5.6.4.1 First Step: Assigning a Value to $C_{\mathrm{in}} V_{\mathrm{h}}$

Using (5.43) and imposing $T_{\mathrm{charge}} < \tau/10$ gives:

$$C_{\mathrm{in}} V_{\mathrm{h}} < \frac{I_{\mathrm{MPP,min}} \tau/10}{2} = 185\ \mu\mathrm{C}. \tag{5.52}$$

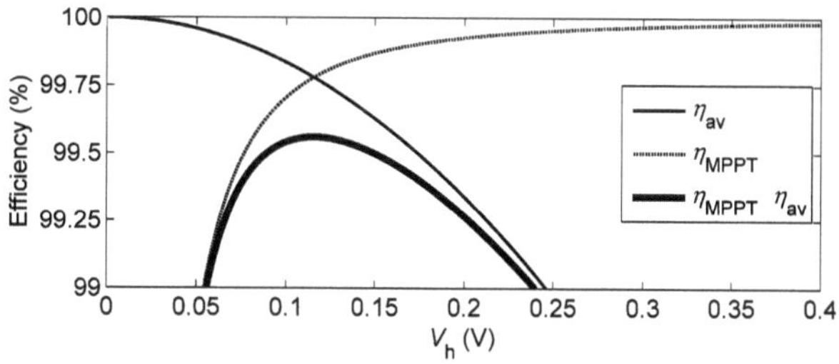

Fig. 5.33 Efficiencies plotted against V_h in the worst-case scenario for the PV module

Since η_T increases for a higher $C_{in}V_h$, a value near the upper bound was chosen: $C_{in}V_h = 100$ μC. Equation (5.52) denotes the tradeoff between the speed of dynamic response and the overall efficiency: the shorter the τ selected, the lower the $C_{in}V_h$ and the resulting η_T.

5.6.4.2 Second Step: Assigning a Value to V_h

Once the product $C_{in}V_h$ was fixed, η_c was observed to be independent of V_h, whereas η_{MPPT} and η_{av} increase and decrease, respectively, with increasing V_h. Thus, the maximum of the product $\eta_{MPPT} \cdot \eta_{av}$ will provide an optimum value of V_h. Figure 5.33 depicts these efficiencies in the worst-case scenario and shows the existence of a maximum near 100 mV. For this PV module, the worst-case scenario corresponds to the minimum P_{MPP} for η_{av} and to the maximum P_{MPP} for η_{MPPT}. So, V_h was assigned a value of 100 mV. For (5.45), considering the information provided in the datasheets of the comparators, $t_{pd1} - 2t_{pd2} + t_{pd3} = 5$ μs was used.

5.6.4.3 Third Step: Determining C_{in} and τ_{INT}

The aforementioned selected values give $C_{in} = 1$ mF. Finally, using (5.51) gives $\tau_{INT} = 88$ ms.

5.6.5 Materials and Methods

We built and implemented a prototype of our proposed MPPT in order to test it with an MSX-005 PV panel. The circuit (Fig. 5.30) was implemented with COTS components. For the dc/dc conversion, a MAX1675 boost converter (PFM control with a current limit of 0.5 A), a 22 μH inductor, and a low ESR 1 mF tantalum input capacitor (C_{in}) were used. LTC1440 and LTC1441 comparators were used to implement CMP1, CMP2 and CMP3. They provide low power (2.1 μA) with a propagation delay of 12 μs. Low-power op amps (*e.g.* OPA2369 and MAX9910) were employed to obtain the limits of the observation window and to amplify the input voltage (v_s). An OPA2369 op amp and a TS5A2066 analog switch were used to implement the low-power S&H. The overall control circuit was powered from the output of the boost converter.

We performed laboratory measurements to assess the static and dynamic performance of the MPPT. In order to achieve reproducible results, the PV panel and the battery illustrated in Fig. 5.30 were substituted with the PV array simulator described in Sect. 5.2 and a voltage source, respectively. The generated P–V curves (see Sect. 5.2) were used to determine the performance of the circuit prototype. Efficiencies were obtained against the whole range of P_{MPP} and at different output voltages. First, η_{MPPT} was calculated by comparing the PV panel power, $P_{\mathrm{s}}(v_{\mathrm{m,eq}})$, with P_{MPP} for each generated I_{SC} of the current source. $P_{\mathrm{s}}(v_{\mathrm{m,eq}})$ was estimated by measuring the average PV panel voltage and obtaining the corresponding power from the experimental P–V curve. Thus, the overall efficiency (η_{T}) was calculated by dividing the average output power ($\overline{P}_{\mathrm{out}}$) by P_{MPP}.$\overline{P}_{\mathrm{out}}$ was estimated from the measurement of the average output voltage and current. A low-ESR 100 mF supercapacitor was added at the output of the circuit to obtain a quasi-dc output current, and therefore, achieve proper measurement of the power efficiency [28]. Finally, the dynamic performance of the MPPT was assessed by programming step-current changes with the current source and then recording the resulting signals with an oscilloscope.

We also performed field measurements to compare the proposed solution with a direct-coupled solution, which we equipped with a low turn-on voltage Schottky diode (BAT47) to improve its efficiency. To avoid any differences arising from use of two PV panels, both circuits were connected alternatively through a reed relay to a single 500 mW PV panel. Once the PV panel was connected to either of the two circuits, a 30 s delay interval was used to let the circuit reach steady-state before the power was recorded (for 1 min). The MPPT had three AA NiMH batteries connected in series at its output, which ensured that the output voltage of the boost converter would always be higher than V_{MPP} at any irradiance level. The direct-coupled circuit had two AA NiMH batteries connected in series at its output, in order to work near the MPP. As explained in Sect. 5.3.1, a direct-coupled circuit for a PV panel with $N = 8$ offers the best efficiencies for the voltage range of two NiMH batteries connected in series.

5.6.6 Experimental Results

5.6.6.1 Laboratory Measurements

Figure 5.34 shows how the measured power consumption (@ 5 V) of the MPPT control circuit increases with the input PV power. Higher PV power translates to a larger I_{SC}, and therefore, a shorter T_{cycle}. Thus, the experimental results agree with (5.19).

Figure 5.35 shows η_{MPPT}. As observed in the figure, η_{MPPT} is greater than 99.6% for the whole power range, demonstrating the efficacy of the proposed MPPT algorithm, and agrees with the corresponding graph of Fig. 5.33 for $V_{\mathrm{h}} = 100$ mV.

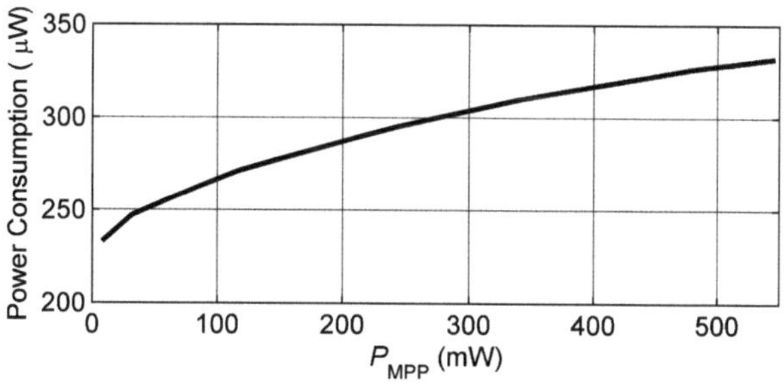

Fig. 5.34 Power consumption (@ 5 V) plotted against programmed P_{MPP} for the MPPT controller circuit

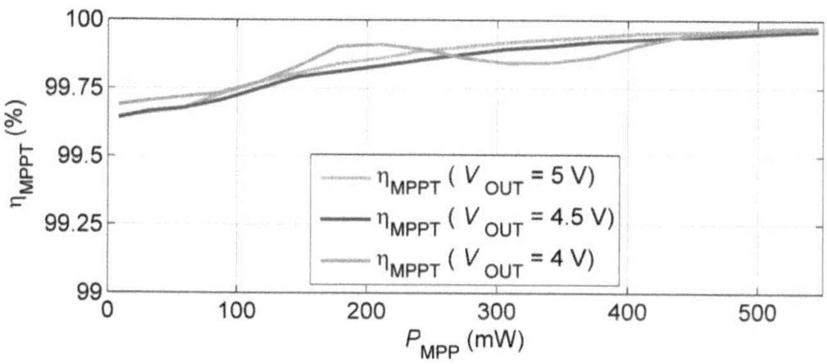

Fig. 5.35 η_{MPPT} plotted against programmed P_{MPP} at three output voltages for the proposed MPPT

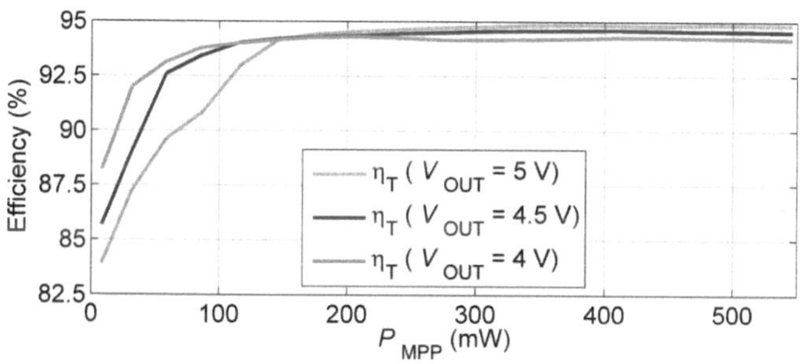

Fig. 5.36 η_T plotted against programmed P_{MPP} at three output voltages for the proposed MPPT

Minor deviations were observed when the output voltage was decreased from 5 V to 4 V. η_{av} was estimated from (5.16) to be higher than 99.8%. This value, together with that of η_{MPPT}, indicates that η_T (shown in Fig. 5.36) is mainly due to η_c (not shown). These results agree with (5.20), which includes the power consumption of the MPPT control circuit and of the dc/dc converter.

Figure 5.37 shows the dynamic response of v_m to step changes in the programmed P_{MPP} value. The time constant (τ) was roughly 1 s, as stated in Sect. 5.6.4.

Figure 5.38 shows the start-up transient of v_{INT}, v_m and v_s obtained when P_{MPP} was increased from 0 mW to 8.2 mW. For this test, the input capacitor (C_{in}) was discharged and the circuitry was powered from the circuit output voltage (5 V) a few seconds prior to the zero time value of the graph. The power change was produced at

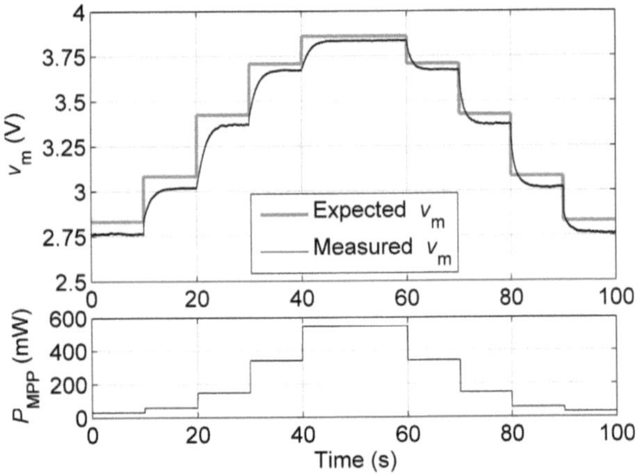

Fig. 5.37 Transient response of v_m to step changes in the programmed P_{MPP} value for the proposed MPPT

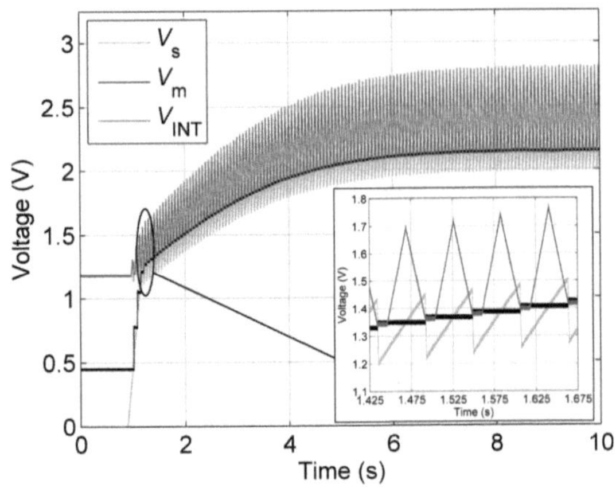

Fig. 5.38 The start-up transient of the proposed MPPT when P_{MPP} was increased from 0 mW to 8.2 mW

$t \approx 1$ s. As observed in the figure, v_m and v_s fast catch v_{INT}, which is initially set to an internal reference value of 1.2 V. Thus, the circuit behaves as a first-order system and v_m tends to the corresponding V_{MPP} value. A detailed look at the temporal evolution of the three represented signals is provided in the inset of Fig. 5.38.

5.6.6.2 Field Measurements

Figure 5.39 shows the power delivered to the batteries during a sunny day for both the proposed MPPT and the direct-coupled circuit. The maximum ambient temperature was 26 °C, and the maximum irradiance over the horizontal surface, 800 W/m^2. The sudden increase in the incoming power at roughly 2:30 p.m. indicates the point at which the PV panel began to receive direct sunlight. The total energy collected by the MPPT was 10.3% higher than that captured by the direct-coupled circuit, even though the measured efficiency for the latter was close to 85% at high irradiance levels.

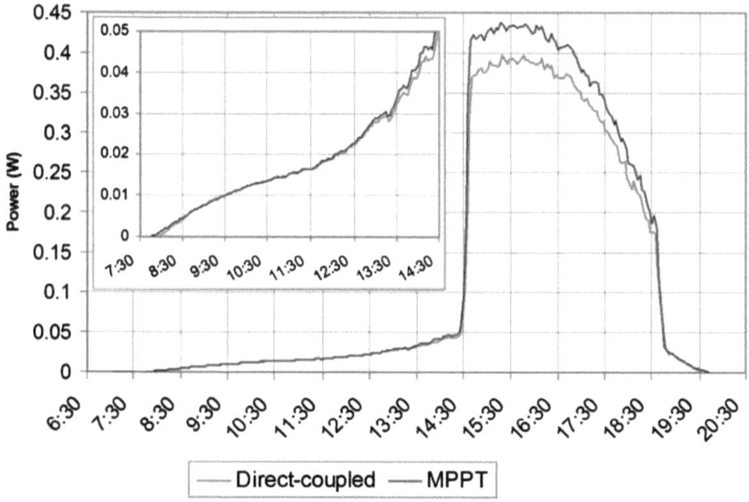

Fig. 5.39 Power delivered to the batteries during a sunny day in a PV panel equipped with either the proposed MPPT or a direct-coupled circuit

As observed in the inset of Fig. 5.39, even at low irradiance levels, from 8:00 a.m. to 2:30 p.m., the power delivered by the MPPT was greater than or equal to that delivered by the direct-coupled one. Firstly, the MPPT circuit started to charge its batteries before the direct-coupled circuit did: the former began charging at a lower V_{OC}, corresponding to the start-up voltage of the MAX1675 (1.1 V), whereas the latter could not begin charging until the V_{OC} of the PV panel was higher than its battery voltage (≈ 2.4 V). As the irradiance increased, the operating point of the PV panel for the direct-coupled circuit approached its MPP. Consequently, from 9:30 a.m. to 1:00 p.m. its efficiency increased to a value close to that of the MPPT, and was mainly limited by the power loss of the diode. The efficiency of the MPPT later increased as the PV power increased (Fig. 5.36), again delivering more power to the batteries than did the direct-coupled circuit.

The MPP voltage of the PV panel increased with decreasing temperature (-16 mV/°C). Thus, the efficiency of the direct-coupled circuit is expected to decrease during cold weather (see Sect. 5.3.1), whereas that of the MPPT is expected to remain high. Therefore, even higher energy gains are expected for the MPPT during these times.

We also concluded that the MPPT outperforms the direct-coupled one in cloudy weather. The two circuits were measured over five consecutive days, and the average power gain was measured hourly (Fig. 5.40). At the end of each day, the diode of the direct-coupled solution stopped the charge of the batteries before the MPPT did, and consequently, the power gains of the MPPT circuit were higher. The batteries of the MPPT were more charged than in the morning, but as expected from Fig. 5.36, its efficiency did not drop significantly. The overall gain was roughly 9%. T_a oscillated between 19 °C and 24 °C. Despite these promising results, a more comprehensive

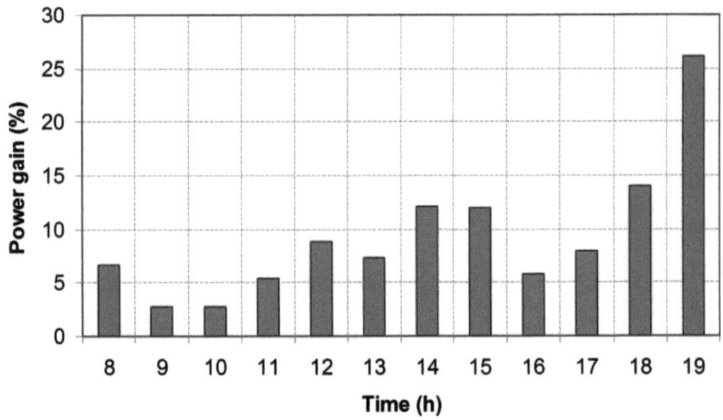

Fig. 5.40 Hourly average power gain over five consecutive days (17th to 21st September, 2009) for the MPPT relative to the direct-coupled circuit

analysis is required which should include experimental measurements taken over an entire year.

5.7 Conclusions

We used a generic model for optical energy transducers (solar cells) to compute I–V and P–V curves and their dependency on optical power and temperature, and to design and optimize several of the energy conditioning circuits proposed in this text. A PV array simulator was implemented using a current source and a dummy PV panel. This simulator was used to perform robust testing of all the proposed energy conditioning circuits.

We analyzed the simplest direct-coupled circuit, in which the PV panel is connected to the storage unit through a diode, at different combinations of series-connected (N) PV cells and secondary batteries, observing efficiencies of roughly 80% to 90% at irradiances higher than 100 W/m^2, when using PV panels with $N = 10$ and either one Li-ion battery or three series-connected NiMH batteries, or PV panels with $N = 7$ and two series-connected NiMH batteries. The efficiency steeply decreases at irradiances lower than 100 W/m^2. For a 500 mW PV panel with $N = 8$ and $2 \times$ NiMH batteries, filed measurements taken on a sunny day revealed an efficiency of approximately 85% for irradiances higher than 50 W/m^2.

We proposed complex energy conditioning circuits based on MPPT controllers and equipped with dc/dc converters employing PFM techniques to raise power efficiency at low power levels. Static performance was assessed according to the overall efficiency (η_T), which in turn was analyzed by subdividing it into three terms: η_{MPPT}, which stems from the MPPT algorithm; η_{av}, generated by the oscillating nature of the input signal of the dc/dc converter when using PFM; and η_c, which comes from the dc/dc converter and MPPT circuitry. Generic expressions were found for

η_{av} and η_c, whereas η_{MPPT} had to be calculated for each MPPT algorithm implemented. High values were generally achieved for η_{av} and η_{MPPT}. Thus, η_T is dominated by η_c—particularly, by the efficiency of the dc/dc converter. The dynamic performance was assessed based on the time response. Generic expressions were devised for the charging and discharging times of the capacitor placed at the input of the dc/dc converter. However, the overall time response also depends on the specific MPPT algorithm used.

We proposed a new open-loop MPPT method, referred as LOCV; firstly, using the generic model for solar cells. The implementations proposed in the literature use the experimental constant factor k for obtaining V_{MPP} from V_{OC}. In contrast, we propose a linear fit between V_{MPP} and V_{OC} that improves the performance, achieving an η_{MPPT} higher than 99.5% within a large range of input irradiances (from $10\ \mathrm{W/m^2}$ to $1000\ \mathrm{W/m^2}$) and at several temperatures (from 0 °C to 45 °C). Experiments with the PV array simulator running at fixed temperature gave η_{MPPT} values greater than 99.9%. A circuit implementation was proposed that achieves a time response shorter than 1 s.

Lastly, we have conceived and tested a new closed-loop MPPT method suitable for low-power PV panels. The proposed technique avoids the measurement of current present in classical methods and techniques. The static and dynamic performances of the proposed MPPT method were theoretically analyzed, and corresponding design criteria were established. The analysis demonstrates that the tracking is highly accurate even for low-power comparators, provided that their propagation delays are matched or the charge cycle is sufficiently long. The dynamic performance was evaluated in function of the time response of v_m, which can be approximated by a first order linear function. Most of the literature precedent has not considered this analysis. A prototype circuit of the new MPPT was implemented using a commercial PFM dc/dc boost converter and low-power components for the MPPT controller. Laboratory measurements of a 500 mW PV panel were taken using the PV array simulator: η_{MPPT} was greater than 99.6%, and η_T, greater than 92%, operating at over 100 mW. This was partly made feasible by the low power consumption of the MPPT controller, which was kept at less than 350 µW. The time response of the tracking circuit was determined to be roughly 1 s. Field measurements were performed to compare the proposed MPPT circuit with a direct-coupled one. Energy gains higher than 10.3% were achieved for the MPPT circuit at an ambient temperature of 26 °C, and even higher gains are expected at lower temperatures.

References

1. M.T. Penella, J. Albesa, M. Gasulla, Powering wireless sensor nodes: Primary batteries versus energy harvesting, in *Proceedings of IEEE International, Instrumentation and Measurement Technology Conference, I2MTC* (2009), pp. 1625–1630
2. F. Lasnier, T.G. Ang, *Photovoltaic Engineering Handbook* (Adam Hilger, Bristol, 1990)
3. N.R. Malik, *Electronic Circuits: Analysis, Simulation and Design* (Prentice Hall, Englewood Cliffs, 1995)
4. A. Luque, S. Hegedus, *Handbook of Photovoltaic Science and Engineering* (2003)

5. BecoSolar, MSX-005 F and MSX-01F polycrystalline OEM modules. Available on: http://www.becosolar.com/DataSheets/MSX-01F%20&%20MSX-005F.pdf. Accessed June 2010
6. V. Raghunathan, A. Kansal, J. Hsu, J. Friedman, M. Srivastava, Design considerations for solar energy harvesting wireless embedded systems, in *Proceedings of Fourth International Symposium on Information Processing in Sensor Networks* (2005), pp. 457–462
7. A. Kansal, J. Hsu, M. Srivastava, V. Raghunathan, Harvesting aware power management for sensor networks, in *Proceedings Design Automation Conference* (2006), pp. 651–656
8. EnOcean, EnOcean: radio switches and sensors without batteries. Available on: http://www.enocean.com/. Accessed October 2, 2006
9. S. Mahlknecht, M. Roetzer, Energy supply considerations for self-sustaining wireless sensor networks, in *Proceedings European Conference on Wireless Sensor Networks, EWSN* (2005), pp. 397–399
10. E. Leder, A. Sutor, R. Lerch, Solar powered low-power sensor module with a radio communication and a user interface, in *Proceedings of IEEE Sensors* (2005), pp. 440–443
11. P. Zhang, C.M. Sadler, S.A. Lyon, M. Martonosi, Hardware design experiences in ZebraNet, in *Proceedings of 2nd International Conference on Embedded Networked Sensor Systems* (2004), pp. 227–238
12. Maxim, Dallas Semiconductor, Harnessing solar power with smart power-conversion techniques. Available on: http://www.maxim-ic.com/app-notes/index.mvp/id/484. Accessed December 2007
13. R.W. Erickson, D. Maksimovic, *Fundamentals of Power Electronics*, 2nd edn. (Kluwer Academic, Norwell, 2001)
14. C. Alippi, C. Galperti, An adaptive system for optimal solar energy harvesting in wireless sensor network nodes. IEEE Trans. on Circuits and Systems I: Regular Papers **55**, 1742–1750 (2008)
15. D. Dondi, A. Bertacchini, D. Brunelli, L. Larcher, L. Benini, Modeling and optimization of a solar energy harvester system for self-powered wireless sensor networks. IEEE Trans. Industrial Electronics **55**, 2759–2766 (2008)
16. S. Hui, T. Chi-Ying, K. Wing-Hung, A micro power management system and maximum output power control for solar energy harvesting applications, in *Proceedings of the 2007 International Symposium on Low Power Electronics and Design* (2007), pp. 298–303
17. T. Esram, P.L. Chapman, Comparison of photovoltaic array maximum power point tracking techniques. IEEE Trans. on Energy Conversion **22**, 439–449 (June 2007)
18. F.I. Simjee, P.H. Chou, Efficient charging of supercapacitors for extended lifetime of wireless sensor nodes. IEEE Trans. Power Electronics **23**, 1526–1536 (2008)
19. V. Salas, E. Olías, A. Barrado, A. Lázaro, Review of the maximum power point tracking algorithms for stand-alone photovoltaic systems. Solar Energy Materials and Solar Cells **90**, 1555–1578 (2006)
20. C. Rodriguez, G.A. J. Amaratunga, Analytic solution to the photovoltaic maximum power point problem. IEEE Trans. on Circuits and Systems I: Regular Papers **54**, 2054–2060 (2007)
21. D. Brunelli, C. Moser, L. Thiele, L. Benini, Design of a solar-harvesting circuit for batteryless embedded systems. IEEE Trans. Circuits and Systems I: Regular Papers **56**, 2519–2528 (2009)
22. L. Dexin, H.C. Pai, Maximizing efficiency of solar-powered systems by load matching, in *Proceedings of the 2004 International Symposium on Low Power Electronics and Design* (2004), pp. 162–167
23. L. Benini, G. De Micheli, Introduction, in *Dynamic Power Management. Design Techniques and CAD Tools* (Kluwer Academic, Norwell, 1998), pp. 1–39
24. M. Gasulla-Forner, O. López-Lapeña, M.T. Penella-López, Method and circuit for the maximum power point tracking of energy transducers, Spanish Patent: P201001280, 2010, in Spanish
25. D. Dondi, D. Brunelli, L. Benini, P. Pavan, A. Bertacchini, L. Larcher, Photovoltaic cell modeling for solar energy powered sensor networks, in *Proceedings of 2nd International Workshop on Advances in Sensors and Interface, IWASI* (2007), pp. 1–6

26. O. Lopez-Lapena, M.T. Penella, M. Gasulla, A new MPPT method for low-power solar energy harvesting. IEEE Trans. on Industrial Electronics **57**, 3129–3138 (2010)
27. M. Gasulla-Forner, O. Lόpez-Lapeña, M.T. Penella-Lόpez, Maximum power point tracking method for solar panels, Spanish Patent: P200902188, 2009, in Spanish
28. M. Day, J. Naik, A practical guide to low power efficiency measurements. Available on: http://www.wirelessnetdesignline.com/showArticle.jhtml;jsessionid=U30WP0GV4MHCBQE1GHOSKHWATMY32JVN?articleID=199203266&queryText=a+and+practical+and+guide+and+to+and+low+and+power+and+efficiency+and+measurements. Accessed April 2007

Chapter 6
Radiofrequency Energy Harvesting

Various reports on harvesting radiofrequency (RF) energy for powering autonomous sensors have been published in the past few years. We considered RF deliberated power distribution over environmental RF energy harvesting because it offers higher power densities and is based on prior knowledge of the RF signal (polarization and orientation). The available data on ultra-high frequency (UHF; 300 MHz to 3 GHz) radiofrequency identification (RFID) systems is invaluable for optimal power source design.

This chapter focuses on proposed methods and circuits for efficient energy management of RF energy harvested for low-power autonomous sensors. The powering distance range can vary from a few centimeters up to several meters. Radiofrequency-powered sensors are assumed to be allocated at a fixed position and distance from the RF power source. Thus, the energy management system can be tailored to a relatively small range of available powers, thereby simplifying design and enhancing circuit efficiency.

The first section of this chapter describes the basic theory behind RF harvesting and reviews some of the literature precedent on its application to autonomous sensors. Then, we present a methodology that we conceived for identifying the best design for a fixed available power at the antenna. Accompanying simulation results are shown. Afterwards, experiments that we performed with an RF energy harvester, implemented using off-the-shelf components, are described and discussed. Finally, the chapter ends with our conclusions on this work, including a few suggesting for conducting future research.

As explained in Chap. 3, we chose a frequency of 868 MHz to avoid interference with the load's communication system, which operates at 2.4 GHz. Nevertheless, the band in 2.4 GHz could be exploited if the power signal was radiated at a time when there are no communications. In this case, a single antenna could be multiplexed for power and communications, although this would entail more complex design. Additionally, as explained in Chap. 3, the powering distance range would be smaller. The band in 433 MHz could also be used, by increasing the power range as well as the antenna dimensions.

M.T. Penella-López, M. Gasulla-Forner, *Powering Autonomous Sensors*,
DOI 10.1007/978-94-007-1573-8_6,

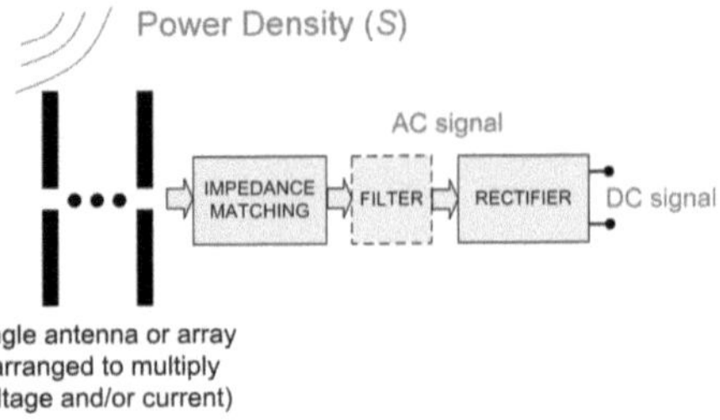

Fig. 6.1 General block diagram of an RF harvester

6.1 Background

Powering systems comprise an RF broadcaster and an RF energy harvester. We concentrate here on the latter. Figure 6.1 shows the building blocks of an RF energy harvester, encompassing an antenna, an impedance matching block, an optional filter and a rectifier. These blocks are described below.

6.1.1 Antenna

Harvesting the energy of an RF signal requires an antenna. Low-frequency RFID systems (mainly 135 kHz and 13.56 MHz) use capacitive or inductive coupling to obtain the power that the chip needs for operation and communication. At these frequencies, the wavelength (λ) of the signal (2222 m for 135 kHz and 22.1 m for 13.56 MHz) is much greater than practical powering distances ($<$ 1 cm for capacitive coupling and $<$ 1 m for inductive coupling) [1]. Thus, the field is in the near-field region and can be considered as an alternating magnetic (for inductive coupling) or electric (for capacitive coupling) field. In the near-field region, magnetic field strength is attenuated according to $1/d^3$ (60 dB per decade of distance [d]). The most widely used antennas at these distances are coils. At a distance of approximately $\lambda/(2\pi)$ the electromagnetic field tends to separate from the antenna and is radiated as an electromagnetic wave. Distances beyond this point are considered as the far-field region, where the electromagnetic wave cannot retroact upon the antenna that generated it (by capacitive or inductive coupling). Nonetheless, the RF-radiated power can still be harvested, using an appropriate resonant structure (antenna). In the far-field region the field strength is attenuated according to $1/d$ (20 dB per decade of d).

As stated in Chap. 3, an antenna can be represented as an ac voltage source with a series impedance (see Fig. 6.2) [2]. In Fig. 6.2, S is the power density, P_{AV} is the available power that the antenna can deliver to a matched load, and A_e is the effective area of the antenna. The series impedance basically comprises the radiation resistance (R_S), the loss resistance (R_{loss}) and a reactive part (X_{ant}). R_S is a virtual resistance that emulates the power used for transmitting/receiving the electromagnetic wave. R_{loss} stems from the actual resistance of the elements that form the antenna (*e.g.* copper) and to dielectric losses. The power dissipated on R_{loss} is

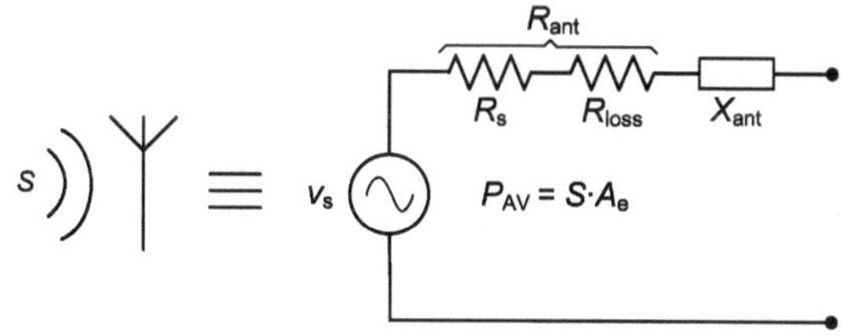

Fig. 6.2 Equivalent electrical model for an antenna [2]

wasted. R_{loss} is low at dc but can increase considerably at high frequencies, due to the *skin effect*, whereby the current is concentrated on the surface of the conductor, consequently increasing its resistance. R_{loss} is typically small compared to R_S at 868 MHz; therefore, we did not consider its effect in this work. In this case, the impedance of the antenna (Z_{ant}) is $R_S + jX_{ant}$. Depending on the antenna design, X_{ant} can be inductive (*e.g.* in loop shape antennas) or capacitive (*e.g.* in patch antennas).

Myriad antennas have been proposed for RFID or RF power harvesters [3–5]. The size of the antenna is inversely proportional to the desired resonance frequency: the smaller the resonant structure, the higher the frequency. This is why printed circuit boards (PCBs) must be carefully designed to avoid undesired resonant surfaces. Despite this frequency dependency, some miniaturization techniques, such as high-dielectric substrates [6] or meandered, bended or tip-loaded structures [7, 8], can be employed to minimize the final size of the system.

The amplitude of the voltage generated on the antenna when matched ($\hat{v}_S$) depends on P_{AV} and R_S:

$$\hat{v}_S = 2\sqrt{2R_S P_{AV}}. \tag{6.1}$$

Thus, at a given P_{AV}, $\hat{v}_S$ increases for a larger R_S. $\hat{v}_S$ must be relatively high in order to reduce the losses from the rectifier. As explained in Chap. 3 and illustrated in Fig. 6.2, P_{AV} depends on A_e and S, and therefore, depends on the antenna characteristics. If greater power is sought, voltages or currents can be added by connecting several antennae (in series and/or in parallel) to form an array; in this case, the antenna's A_e is increased at the expense of a larger physical area [9, 10]. A single rectifying circuit for the whole array reduces the number of rectifying elements but can complicate array design. An antenna with a rectifying element and the necessary matching elements is called a *rectenna*. Rectennas can also be connected in series or parallel in order to add current or voltage; again, at the expense of greater effective area (and physical dimensions). Efficiency depends on input power, and for microwatt-level inputs is 20% maximal [9].

6.1.2 Impedance Matching

The transmitted power reaches its maximum when the antenna sees at its output an impedance that is the conjugate of its own impedance: $Z^*_{ant} = R_S - jX_{ant}$ (see

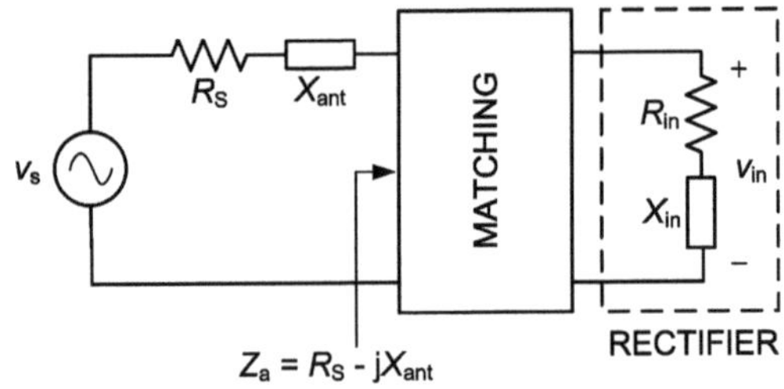

Fig. 6.3 Electrical model of an antenna, an impedance matching network, and the equivalent input impedance of the rectifier

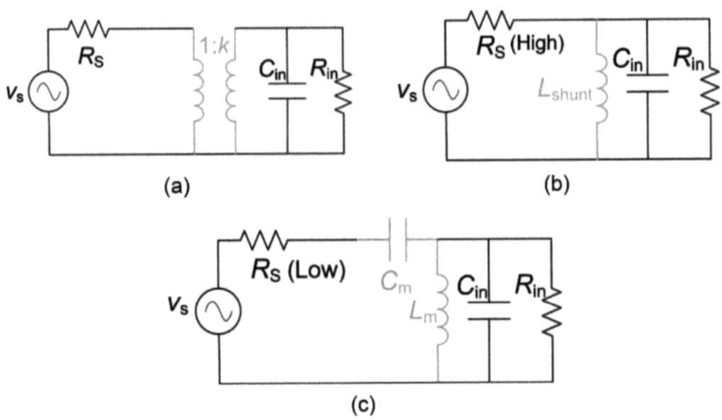

Fig. 6.4 Typical matching network circuits for RFID or RF harvesting circuits: (**a**) transformer; (**b**) shunt inductor; and (**c**) *LC* network. The matching network elements are shown in *grey*

Z_a in Fig. 6.3). In Fig. 6.3 $Z_{in} = R_{in} + jX_{in}$ is the impedance of the rectifier and ensuing load, and R_{in} and X_{in} are the real part and imaginary part of this impedance, respectively. This procedure is known as *impedance matching*. A properly matched design is as important as a well-designed antenna or rectifier. Impedance matching networks are usually made with reactive components (coils or capacitors) that are ideally not dissipative.

Numerous matching configurations are available. The choice of configuration depends on the number of elements used for the matching and the way in which they are positioned. Nonetheless, to date, three main circuits have been proposed for RFID or RF harvesting: *a transformer* (Fig. 6.4(a)), *a shunt inductor* (Fig. 6.4(b)) and an *LC network* (Fig. 6.4(c)). In Fig. 6.4 the antenna is modeled as a voltage source with a real impedance and the rectifier is modeled with a resistance (R_{in}) and a parallel capacitor (C_{in}). If the antenna is not purely real, its reactance must be considered for the matching. The rectifier model is further explained in Sect. 6.1.4.

Using a transformer as matching network (Fig. 6.4(a)) boosts the input voltage by the factor k, which is the ratio of the transformer's secondary voltage to its primary voltage. This solution is cost prohibitive in RFID applications whose cost is limited to a few cents, but it can be considered for autonomous sensors [11]. However, we did not further consider it here.

The shunt inductor matching network is widely used in RFID transponders [2]. To match the antenna and rectifier impedances, L_{shunt} must resonate with the C_{in} of the rectifier. Therefore, their impedances must be equal at the working frequency:

$$L_{\text{shunt}} = \frac{1}{\omega_r^2 C_{\text{in}}}, \tag{6.2}$$

whereby ω_r is the angular frequency at resonance. Thus, as C_{in} is increased, L_{shunt} must be reduced. Relatively high values of C_{in} may require a very small value of L_{shunt}, which can be impractical. If the reactive parts are well matched, the input voltage to the rectifier will come from the voltage divider between R_S and R_{in}, which is the maximum voltage obtainable with this circuit. The most common response to the increased voltage generated on the antenna that results from increased R_S (see (6.1)) is to increase the input voltage of the rectifier (v_{in}) and reduce the losses of the ensuing rectifier. This is accomplished by implementing high radiation resistance antennas (*e.g.* a folded dipole has a radiation resistance of roughly 300 Ω [12]).

The quality factor (Q) of a matching network is the ratio of the resonance frequency (f_r) to the circuit bandwidth at −3 dB (BW); thus, the higher the Q, the smaller BW. If BW is too narrow, a small deviation in f_r will cause a massive power drop at the rectifier input; therefore, BW should be kept sufficiently high. The circuit with the shunt inductor matching network is an *RLC* parallel circuit with a resistance that is the parallel association of R_S and R_{in} ($R_P = R_S \| R_{\text{in}}$). Therefore:

$$Q = \frac{f_r}{\text{BW}} = R_P C_{\text{in}} \omega_r = \frac{R_P}{\omega_r L_{\text{shunt}}}. \tag{6.3}$$

Equation (6.3) indicates that a high C_{in} (or low L_{shunt}) will result in a high Q, and consequently, a narrow BW. Thus, C_{in} must be kept low enough to obtain a reasonable BW. Furthermore, a high value of R_S increases both v_{in} and Q but decreases BW.

The *LC* matching network boosts the antenna voltage. It is advantageous because it provides higher peak voltage to the rectifier and offers a comparatively better efficiency, especially when P_{AV} (and therefore, $\hat{v}_S$) is low. This circuit is also advantageous in antennas with low R_S that also present a lower $\hat{v}_S$. To obtain a resonant circuit, the impedance faced by the matching network must be the conjugate of the antenna impedance. Thus, equating the real and imaginary parts of both impedances generates two equations that enable calculation of the values of L_m and C_m [13]:

$$L_m = \frac{R_{\text{in}}}{\omega_r} \frac{1}{R_{\text{in}} C_{\text{in}} \omega_r + \sqrt{\frac{R_{\text{in}} - R_S}{R_S}}}, \tag{6.4}$$

$$C_m = \frac{1}{R_S \omega_r} \sqrt{\frac{R_S}{R_{\text{in}} - R_S}}. \tag{6.5}$$

Equations (6.4) and (6.5) reveal that L_m depends on the value of C_{in}, but that C_m does not: it only depends on R_S, R_{in} and ω_r. As with the shunt inductor matching network, L_m decreases as C_{in} increases. Therefore, C_{in} must be kept low enough to obtain a feasible value of L_m. Using (6.4) and (6.5) to compute the relationship between v_{in} and v_s at resonance affords:

$$H(\omega_r) = \frac{v_{in}(\omega_r)}{v_s(\omega_r)} = \frac{1}{2}\sqrt{\frac{R_{in}}{R_S}}, \tag{6.6}$$

and Q results in [13]:

$$Q = \frac{1}{2}\left[R_{in}C_{in}\omega_r + \sqrt{\frac{R_{in} - R_S}{R_S}}\right], \tag{6.7}$$

which shows that when $R_{in} > R_S$, the voltage will be boosted and the circuit becomes more selective (higher Q and more narrow bandwidth). The value of R_{in} required to boost the voltage to a given value can be determined from (6.6). Equation (6.5) can then be used to calculate C_m. The value of L_m must be found by trial and error, as C_{in} is not accurately known. The parasitic resistance of the matching elements will ultimately limit the overall Q of the circuit, and consequently, will limit the boosting factor.

Some antennas are custom designed to be power matched to the rectifying stage for a certain frequency and power range [9]. The matching elements can be implemented on the PCB design or can be off-the-shelf components. Printed circuit board traces can be designed to exhibit a definite inductive or capacitive value at the desired frequency, which lowers the cost of the components.

6.1.3 Filters

Low-pass filters are used to prevent the harmonics generated by the non-linearity of the rectifying diodes from flowing back to the antenna, as these harmonics can reduce the peak amplitude of v_{in} ($\hat{v}_{in}$) [14], and consequently, diminish efficiency. Filters demand two impedance matching circuits: one between the antenna and the filter, and one between the filter and the rectifier. Ultra-high frequency (UHF) RFID tags are not typically equipped with filters.

6.1.4 Rectifier

Rectifier circuits provide a dc output voltage at the ensuing load. There are three main options for the rectifier:

- a diode (which, together with the antenna, forms a rectenna) [9] (Fig. 6.5(a));
- a bridge of diodes (or diode-connected transistors) [15] (Fig. 6.5(b));
- a voltage rectifier multiplier [2] (Fig. 6.6(a)).

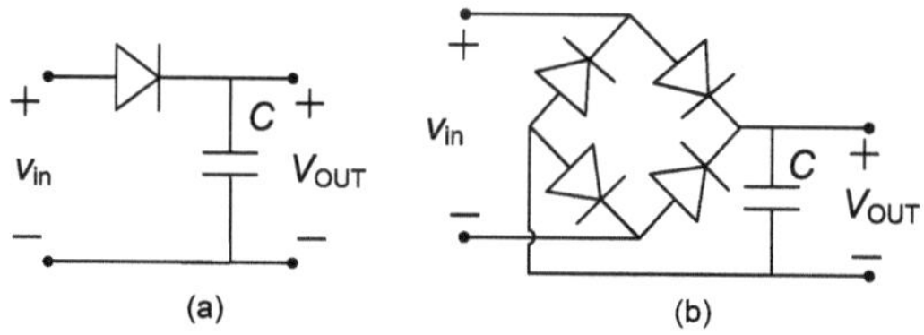

Fig. 6.5 (**a**) Single diode and (**b**) diode bridge rectifiers

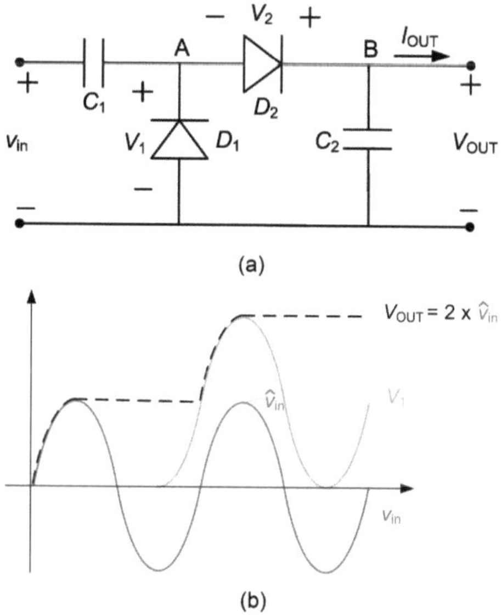

Fig. 6.6 (**a**) A single stage of a voltage multiplier rectifier and (**b**) waveforms during the transient

All of these circuits are broadband (*i.e.* they exhibit the same rectifying characteristics in a broad band of frequencies). For kilohertz or megahertz frequencies, p-n diodes and transistors are used as rectifiers, whereas for microwave frequencies (> 1 GHz), Schottky diodes (GaAs or Si type) are preferred, because they have shorter transit times. The rectification performance of the diode depends on its saturation current, its junction capacitance and its conduction resistance. A large saturation current is sought, because it leads to a low forward voltage drop.

The diode (Fig. 6.5(a)) and the diode bridge (Fig. 6.5(b)) provide an output dc voltage to the load (V_{OUT}) whose amplitude is lower than that of the incoming signal. These circuits are explained in the literature and are not further described in this section. The *voltage rectifier multiplier*, as its name indicates, multiplies the peak amplitude of the incoming signal. At long distances (low P_{AV}, and consequently, low $\hat{v}_S$; see (6.1)), the dc voltage level is not high enough to power an electronic circuit, so the voltage rectifier multiplier appears to be the best solution [2]. The remaining part of this section is dedicated to describing this circuit and reporting on analyses that we performed of it.

The voltage rectifier multiplier is sometimes called a *Dickson charge pump*, but it is actually a modified version of this pump [11]. The behavior of an ideal single-stage rectifier multiplier, such as that shown in Fig. 6.6(a), is explained below. We

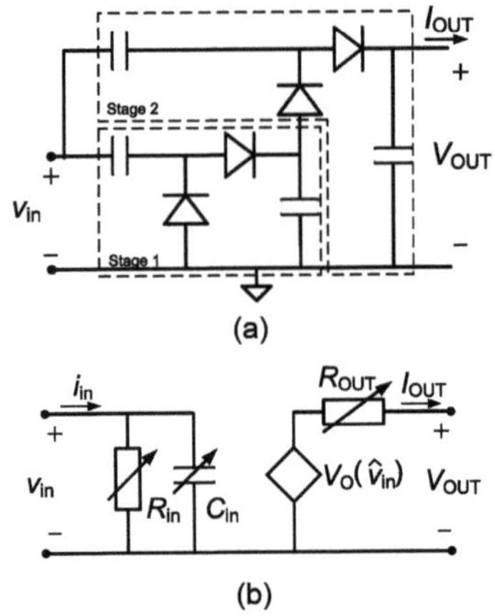

Fig. 6.7 (**a**) Two-stage voltage multiplier rectifier and (**b**) equivalent rectifier circuit model

assume ideal diodes (*i.e.* 0 V threshold voltage, no reverse current, and zero conduction resistance) and that v_{in} is a sinusoidal wave.

Diode D_1 and capacitor C_1 establish a dc reference for the voltage at point A (Fig. 6.6(a)). D_1 conducts whenever the voltage at point A is negative, and then builds an average charge at C_1 that prevents the voltage at point A from ever going negative. If all the elements are ideal, the voltage at this point is $\hat{v}_{in} + v_{in}$ (V_1 in Fig. 6.6(b)). D_2 rectifies the signal at point A and capacitor C_2 holds the output voltage (V_{OUT}) at the peak value of V_1. Thus, the open circuit voltage (OCV) of the rectifier is a dc voltage with a value $2\hat{v}_{in}$. In steady-state operation, I_{OUT} is drained from C_2, and this capacitor is recharged when V_1 is higher than V_{OUT} (dashed line in Fig. 6.6(b)). $\hat{v}_{in}$ can be in the millivolt level, so in order to obtain higher output voltages, the circuit can be cascaded by using N stages (Fig. 6.7(a) shows a two stage rectifier multiplier). In the case of an N-stage rectifier, V_{OUT} in open circuit is:

$$V_{OUT} = 2N\hat{v}_{in}. \tag{6.8}$$

In a lossless rectifier, the input power (P_{in}) equals the output power (P_{OUT}), and R_{in} can be calculated from the resistance connected at the output (R_{Load}) as:

$$R_{in} = \frac{\hat{v}_{in}^2}{2P_{in}} = \frac{V_{OUT}^2}{P_{OUT}} \frac{1}{8N^2} = \frac{R_{Load}}{8N^2}. \tag{6.9}$$

Thus, for a fixed R_{Load}, as N increases, R_{in} decreases. Otherwise, for a matched value of $R_{in} = R_{ant}$, the optimum value of R_{Load} increases as N increases.

When considering the threshold voltage of the diodes (V_T), the OCV is:

$$V_{OUT} = 2N(\hat{v}_{in} - V_T). \tag{6.10}$$

Figure 6.7(b) shows an equivalent model of a rectifier that was exhaustedly analyzed in [11] and [16]. The input impedance comprises a resistance (R_{in}) in parallel with a capacitor (C_{in}) that models the capacitance presented by the layout and the diodes. For optimal performance this impedance must be properly matched to the antenna. The output is modeled as a voltage source (V_O) controlled by $\hat{v}_{in}$, with a serial output resistance (R_{OUT}). R_{OUT} models the losses of the rectifier, which are mainly due to the increasing voltage drop at the diodes when the output current is increased.

An accurate mathematical analysis of the real rectifier is quite complex, but simulations can be used to facilitate rectifier design. Furthermore, the whole system (antenna, matching network and rectifier) must be jointly analyzed in order to obtain meaningful results.

The diodes of the rectifier can also be implemented using CMOS transistors connected as diodes, which is the case for most RFID tags [12]. Using CMOS transistors as diodes implies a design tradeoff among the reverse current, the direct current (threshold voltage and current slope for the rectifying stage), and the junction and parasitic capacitances: improving one of these may compromise another. For example, to increase the direct current slope (which causes the efficiency to increase), the width of the device must be increased, which also results in higher reverse current, and therefore, in losses (efficiency decreases). Using rectifying elements with low threshold voltage and low reverse current will increase the sensitivity of the rectifier (*i.e.* the minimum input voltage at which the rectifier can work under given load requirements). Additionally, the performance of the rectifier at low values of $\bar{P}_{\text{in}}$ is greatly affected by the threshold voltage of the diodes.

To overcome the technological issues of traditional CMOS, researchers have proposed several rectifier designs. For example, to obtain transistors with very low threshold voltage, Curty et al. employed the silicon-on-sapphire (SOS) CMOS process [16], and Karthaus and Fischer used silicon-titanium Schottky diodes [17]. Others have avoided using these relatively expensive processes by biasing the voltage at the gate of the transistor, which results in almost zero threshold voltage transistors [8, 18]. Radiofrequency ID tags usually include the required number of stages that give the desired output performance at the expected minimum $\bar{P}_{\text{in}}$.

6.1.5 Post-rectification Energy Conditioning

To boost the output voltage of the rectifier whenever it is too low, some step-up technologies that have been applied to other energy harvesters (*e.g.* thermal harvesters) can be used. Commercial step-up converters have start-up voltages around 0.7 V. In [19], several circuits with start-up voltages as low as 0.3 V were proposed. Furthermore, a custom dc/dc converter working with a battery at its output could start at lower input voltages. Regardless of the implementation, these circuits must use very low-power techniques.

Paing et al. harnessed a dc/dc converter to boost the rectified output voltage of a rectenna to achieve the desired output voltage at long distances [20]. They also implemented an MPPT solution based on the assumption that the rectenna performs better with a fixed resistance at its output. They set the dc/dc converter to work as a resistor emulator. The required value of the emulated resistance is firstly determined and then the dc/dc converter is programmed to present this value at its input, so an initial characterization process is necessary. They used an open loop control strategy that keeps the control power consumption very low ($\sim 20\ \mu W$). The reported overall efficiency was of 16.7% for a power density of $70\ \mu W/cm^2$ at the antenna (P_{AV} was around 2.5 mW) when connecting a thin-film lithium battery (4.15 V) at the output of the dc/dc converter.

In addition to the method proposed in [20], some of the techniques used for optical energy harvesting (see Chap. 5) can be also applied to RF harvesting. To achieve a net power gain, the power consumption of the energy conditioning circuit must be in the microwatt level or below. If well designed, low power microcontrollers can confer some intelligence to an MPP tracking system at the cost of few microwatts.

6.1.6 *Radiofrequency Energy Harvesting for Autonomous Sensors*

Several works deal with the design of a whole sensor platform powered from the incoming RF signal [18, 21–23] or attempt to recharge a microbattery [24]. In [18], the authors achieved 1.5 V at -12 dBm input power, leading to a maximum powering distance of 10 m in the 915 MHz band, when emitting 4 W of EIRP, whereas in [21] the authors report that the WISP platform can deliver a voltage of approximately 1 V at a distance of 4 m in the same band. In [24], the authors reached output voltages below 0.5 V at $P_{AV} = 0$ dBm. The company Power Cast has announced a wireless sensor powered by RF energy that delivers roughly 10 μA at an output voltage of 1.1 V and $P_{AV} = -10$ dBm [25].

6.2 Analysis

This section describes simulations that we performed on different circuit topologies:

- A voltage multiplier rectifier (Fig. 6.7(a)) with a different number of stages (1 to 5), at three values of P_{AV} (10 dBm, 0 dBm and -10 dBm), with an antenna of $R_S = 300\ \Omega$ and a shunt inductor matching network. A single diode rectifier was also considered, at $P_{AV} = 10$ dBm.
- A one-stage multiplier rectifier, at two values of P_{AV} (0 dBm and -10 dBm), with an antenna of $R_S = 50\ \Omega$ and an LC matching network.

A 300 Ω antenna can be practically implemented with a folded dipole [26]. Higher radiation resistances (up to 2 kΩ) can be reached with some variations to the folded dipole, but high R_S antennas can be problematic (see Sect. 6.1.2 and [2]), due to the finite quality factor of the matching elements and the input parasitics of the rectifier.

Figure 6.8 shows the general block diagram of the circuit we used for the simulations. For the antenna, we used an RF port with a serial resistance of the desired value (300 Ω or 50 Ω). The load comprised a 1 nF capacitor in parallel with a resistance. The value of this resistance was swept across a suitable range to obtain the

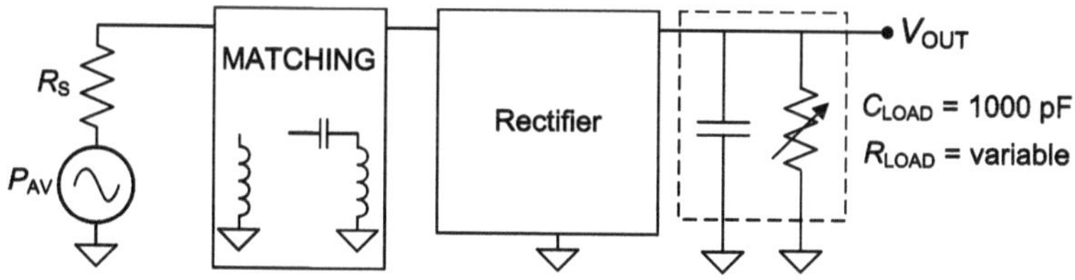

Fig. 6.8 General circuit schematic used for simulations

total efficiency of the circuit (η_o) versus the output voltage of the rectifier (V_{OUT}). η_o was computed as:

$$\eta_o = \frac{P_{DC}}{P_{AV}}, \tag{6.11}$$

whereby P_{DC} is the rectified output power and is equal to $V_{OUT} I_{OUT}$.

We ran the simulations using the program Advanced Design System (ADS), from Agilent. It enables circuit (schematic) simulations and layout or electromagnetic (momentum) simulations, or a combination of both, which should offer good agreement with real laboratory experiments. Nonetheless, for this work we only used the circuit simulation. A thorough simulation analysis is essential for saving time and effort when implementing the real circuit.

The design methodology comprised the following steps. For the circuits with a shunt inductor matching network, a suitable L_{shunt} was found by trial and error for each specific value of P_{AV}. Then, η_o was plotted against V_{OUT} at each value of P_{AV}. For the circuits with an LC matching network, a value of C_m was fixed, and L_m was obtained by trial and error. Then, η_o was again plotted against V_{OUT}.

This section ends with a discussion on the use of appropriate storage units for the analyzed circuits at each specific value of P_{AV}.

6.2.1 Rectifier

For the multiplier rectifier, we assembled the basic stage (Fig. 6.9) using HSMS-2822 diodes (Avago Technologies) with two Schottky diodes connected in series. Each diode has a 340 mV forward voltage at 1 mA current. The series configuration enabled mounting of one stage with a single diode chip. These diodes are suitable for frequencies of up to 1.5 GHz and forward current of up to 20 mA and have a junction capacitance of roughly 0.7 pF. Signal detection diodes that have very small forward voltages are not suitable for this application because they cannot stand input powers greater than −20 dBm. The intermediate capacitors were set to 12 pF. In order to not further increase C_{in}, the impedance of these capacitors must be kept as low as possible relative to the diode capacitance [16].

Figure 6.9(a) shows a single stage of the rectifier. Several stages can be cascaded by connecting the *output* of one stage to the *diode reference* of the next stage. The *RF input* is connected to the RF source (*i.e.* the antenna).

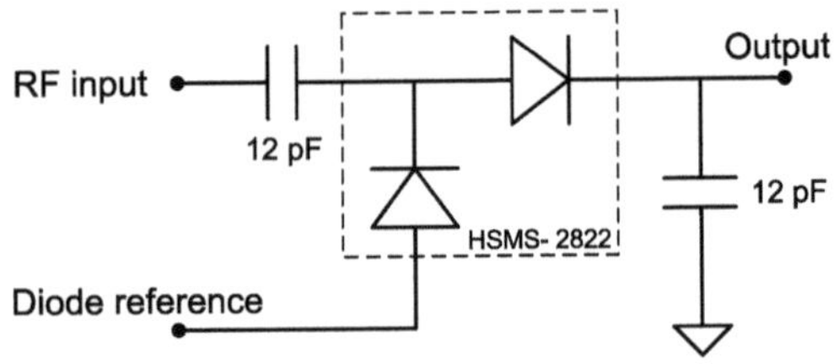

Fig. 6.9 Circuit schematic of one of the stages used for the rectifier

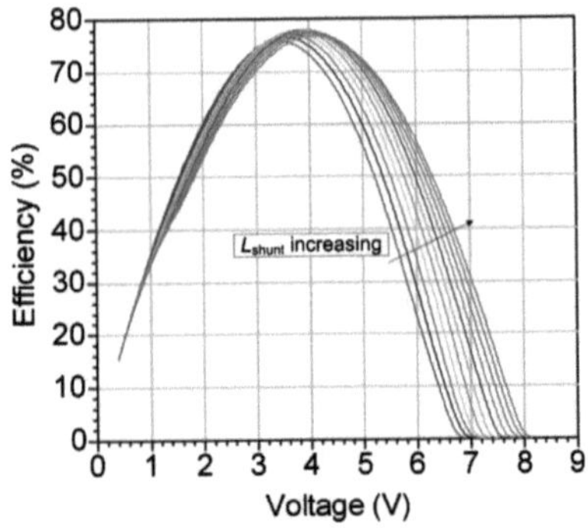

Fig. 6.10 Efficiency plotted against V_{OUT} for a one stage rectifier at $P_{AV} = 10$ dBm and when L_{shunt} was varied in 1 nH steps from 30 nH to 40 nH

For the single diode rectifier (Fig. 6.5(a)), an HSMS-2820 and an output capacitance of 1 nF were used. The HSMS-2820 includes just one diode of the same type as those included in the HSMS-2822.

6.2.2 *Simulations*

6.2.2.1 $P_{AV} = 10$ dBm

Figure 6.10 shows η_o plotted against V_{OUT} for a single-stage rectifier at $P_{AV} = 10$ dBm and at values of L_{shunt} ranging from 30 nH to 40 nH. As with solar cells, there is a voltage value where the efficiency is maximum (V_{MPP}): in this case, roughly 4 V. As observed in the figure, the maximum efficiency is quite high ($> 75\%$) and does not significantly change among the different values of L_{shunt}. This is mainly due to the high voltage present at the rectifier input ($\hat{v}_{in} = 2.45$ V, provided that the antenna is properly matched), which keeps the relative losses at the diodes low. An intermediate value for L_{shunt} (35 nH), which provided good overall efficiency, was chosen. Efficiencies sharply decrease when the output voltage shifts from V_{MPP}. For the higher voltages, efficiency worsens for decreasing values of L_{shunt}. Conversely, smaller voltages lead to similar efficiencies for the entire range of L_{shunt} values.

Table 6.1 lists the L_{shunt} value found for each N. As expected from Sects. 6.1.2 and 6.1.4, L_{shunt} decreases with increasing N, due to increasing C_{in}.

Figure 6.11 shows efficiency plotted against V_{OUT} for the different stage rectifiers with L_{shunt} values shown in Table 6.1. The efficiencies at the MPP are higher

Table 6.1 L_{shunt} for each N at $P_{AV} = 10$ dBm

N	L_{shunt} (nH)
1	35.0
2	20.5
3	14.0
4	11.0
5	8.5

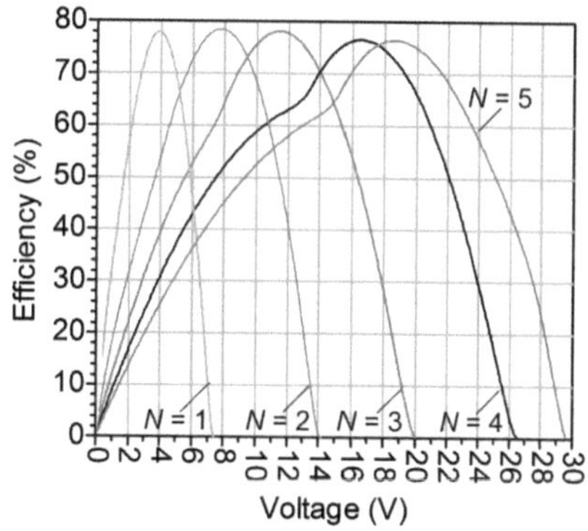

Fig. 6.11 Efficiency plotted against V_{OUT} for stage rectifiers with N values of 1 to 5 and at $P_{AV} = 10$ dBm

than 75%. A single-stage rectifier offers the best efficiency for V_{OUT} up to 5 V, which includes the voltage range of the storage units typically used for autonomous sensors (one Li-ion battery, or two or three series-connected NiMH batteries). As N increases, V_{MPP} increases to values far beyond the supply voltage range of the autonomous sensor. Thus, operation at V_{MPP} requires a step-down dc/dc converter positioned after the rectifier or the storage unit. As commented in Chap. 5, the efficiency of step-down converters is lowered when their input voltage is well above their output voltage [27]. Therefore, the single-stage rectifier is the best solution for this P_{AV}, as it offers the best efficiency at V_{OUT} below 5 V with the lowest circuit complexity and component cost.

Given that the voltage generated on the antenna is already high ($\hat{v}_S = 4.89$ V), a single diode rectifier (Fig. 6.5(a)) can also offer good efficiency. Figure 6.12 shows the simulations for this circuit when connected to a 300 Ω antenna with $L_{shunt} =$ 64 nH. As observed in the figure, the maximum efficiency is slightly lower than with a single-stage rectifier; however, the efficiency is higher for $V_{OUT} < 2$ V. Thus, this circuit offers a better solution if a single NiMH battery is used as storage unit. Nevertheless, a further analysis of this circuit would be required, as we had expected a similar maximum efficiency to that shown in Fig. 6.11.

6.2.2.2 $P_{AV} = 0$ dBm

Figure 6.13 illustrates how efficiency changed for a single-stage rectifier at $P_{AV} =$ 0 dBm when L_{shunt} ranged from 25 nH to 30 nH. As observed in the figure, the

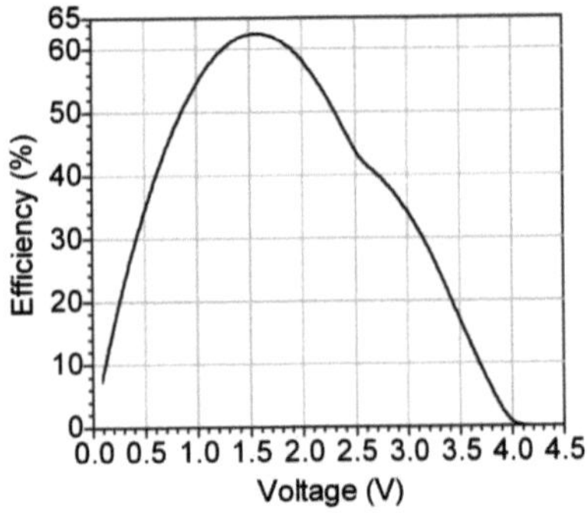

Fig. 6.12 Efficiency plotted against V_{OUT} for a single diode connected as in Fig. 6.5(a) with $R_S = 300\ \Omega$, $L_{shunt} = 64$ nH and $P_{AV} = 10$ dBm

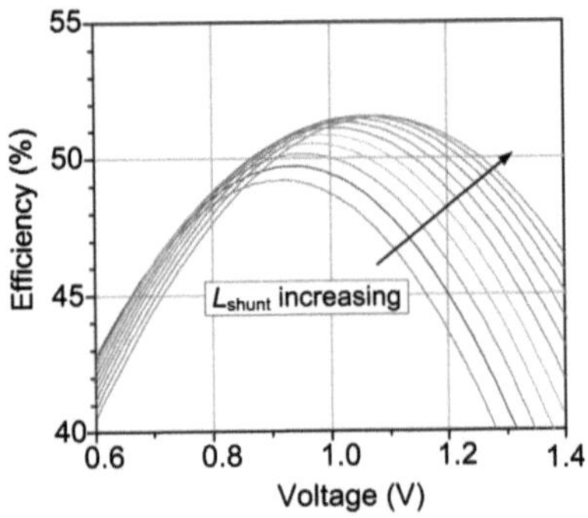

Fig. 6.13 Efficiency plotted against voltage for a one-stage rectifier at $P_{AV} = 0$ dBm and when L_{shunt} was ranged from 25 nH to 30 nH

Table 6.2 L_{shunt} for each N at $P_{AV} = 0$ dBm

N	L_{shunt} (nH)
1	29.5
2	16.0
3	10.5
4	8.0
5	6.5

maximum efficiency is roughly 52%, lower than that for $P_{AV} = 10$ dBm. This can be explained by the fact that in the former case, the voltage at the rectifier input is lower ($\hat{v}_{in} = 745$ mV), meaning that the relative losses at the diodes are higher. Likewise, V_{MPP} is also lower, at values near 1 V. As with the case of $P_{AV} = 10$ dBm, efficiency beyond the MPP is strongly affected by the value of L_{shunt}.

We sought the optimum value of L_{shunt} for each N. The results are listed in Table 6.2. The L_{shunt} values are smaller than those found at $P_{AV} = 10$ dBm, which is probably due to the increase in C_{in} as P_{AV} decreased.

A curve of efficiency plotted against V_{OUT} is shown in Fig. 6.14. The single-stage rectifier is the best suited circuit for charging a single NiMH battery. The value of V_{MPP} is near the voltage range of a single NiMH battery ($1.2\text{ V} < V_{OUT} < 1.5\text{ V}$). The efficiency of a single diode solution (not shown) will be much worse at this

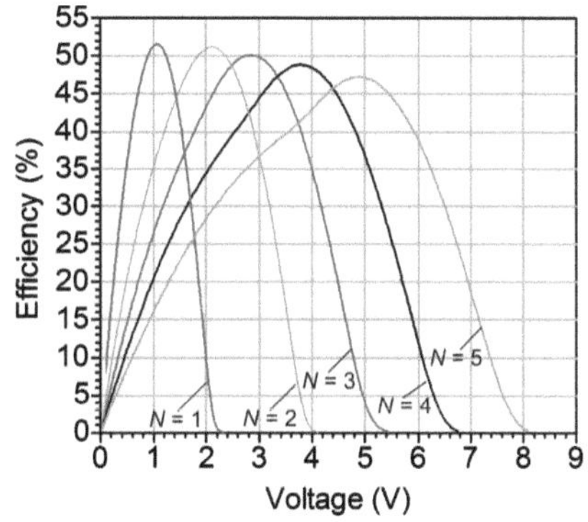

Fig. 6.14 Efficiency plotted against V_{OUT} for an N-stage rectifier at $P_{AV} = 0$ dBm

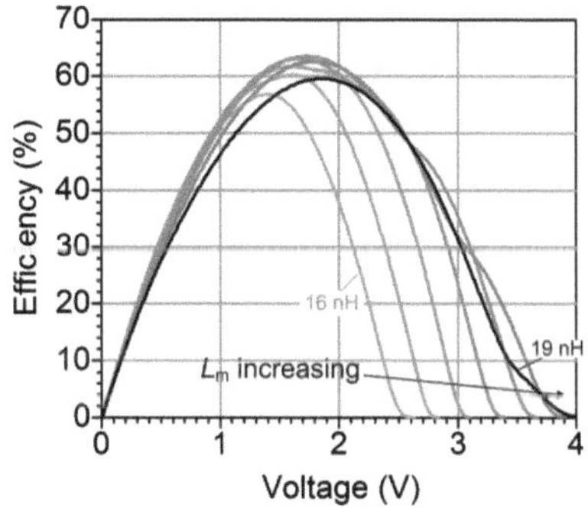

Fig. 6.15 Efficiency plotted against output voltage for a one-stage rectifier at $R_S = 50\ \Omega$ and $P_{AV} = 0$ dBm. For the LC matching network, $C_m = 1$ pF and different values of L_m (from 16 nH to 19 nH) were used

voltage range. Rectifiers with $N = 2$ or $N = 4$ are optimal for directly charging two NiMH or three NiMH batteries (or one Li-ion battery), respectively.

We also considered the use of an LC matching network with a single-stage rectifier and an antenna with $R_S = 50\ \Omega$. In this case, $\hat{v}_S = 0.63$ V, which is less than half the value achieved with $R_S = 300\ \Omega$. C_m was fixed at 1 pF, which leads (from (6.5)) to an optimal matching R_{in} of 722 Ω and (from (6.6)) to a maximum voltage boost of 1.9 times. The results for different values of L_m (from 16 nH to 19 nH) and $C_m = 1$ pF are shown in Fig. 6.15. As observed in the figure, the maximum efficiency was higher than in the case corresponding to Fig. 6.14, even though the value of R_S was smaller. This circuit, when directly connected to one or two series-connected NiMH batteries, exhibited η_o of 40% to 60% over the entire battery voltage range; the best overall η_o in this range was obtained at $L_m = 18$ nH. Thus, the single-stage rectifier with an LC matching network and $R_S = 50\ \Omega$ outperformed the system corresponding to Fig. 6.15 for $N = 1$ or $N = 2$, which were the best solutions for the voltage range of one or two NiMH batteries, respectively.

6.2.2.3 $P_{AV} = -10$ dBm

In this case, both the expected P_{OUT} and V_{OUT} were low. Therefore, we tried to maximize efficiency at $V_{OUT} > 1$ V to avoid having to place additional power

Table 6.3 L_{shunt} for each N at $P_{AV} = -10$ dBm

N	L_{shunt} (nH)
1	25.3
2	14.7
3	9.5
4	7.1
5	5.6

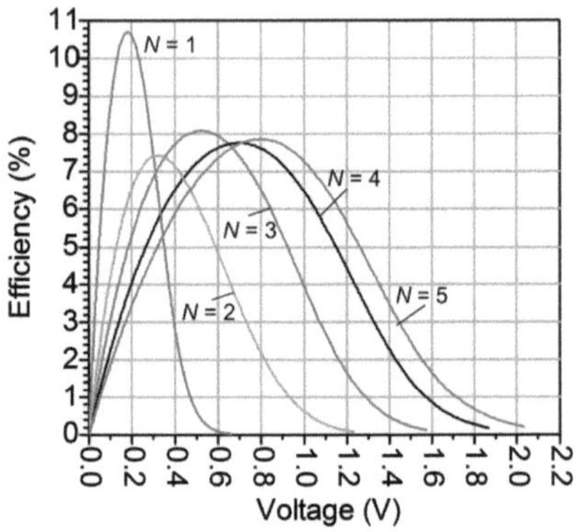

Fig. 6.16 Efficiency plotted against V_{OUT} for an N-stage rectifier at $P_{AV} = -10$ dBm

management circuits at the output. The values of L_{shunt} that maximized efficiency at $V_{OUT} > 1$ V for each value of N are listed in Table 6.3. For the single-stage rectifier, a value of 1 V could not be reached. The L_{shunt} that enabled maximum efficiency was used.

The efficiency of the N-stage rectifier at different values of N is plotted in Fig. 6.16. In this case, the efficiencies were rather low, as $\hat{v}_{in}$ was 245 mV. When using a one-stage rectifier, the harvested power could only be stored on a capacitor, and a subsequent boost converter had to be used to raise the voltage and power the autonomous sensor. However, the small values of V_{MPP} and of P_{OUT} ($\sim 10\,\mu$W) do not favor the use of a boost converter. At this P_{AV}, the best solution based on a 300 Ω antenna uses a five-stage rectifier. Nonetheless, the efficiency at $V_{OUT} = 1$ V (one discharged NiMH battery) is very poor (7.2%) and I_{OUT} is 7.2 μA. For this case, using a higher number of stages in the rectifier might be advantageous.

Figure 6.17 shows the efficiency when using an LC matching network with an antenna with $R_S = 50$ Ω and a single-stage rectifier. For the matching network, $C_m = 0.5$ pF (see Sect. 6.1.2) and different values of L_m (from 20 nH to 22 nH) were used. The resulting rectifying voltage boost was 3.7 times, which enabled efficiencies of 40%, even though $\hat{v}_S$ was as low as 0.2 V. However, in this case, efficiency is more sensitive to variations in L_m, because of the higher value of Q (and thus, the smaller BW). Furthermore, efficiency sharply decreases when shifting from V_{MPP} (roughly 1 V). This effect did not occur when a multistage rectifier with $R_S = 300$ Ω and shunt inductor matching network (grey curve in Fig. 6.17) were used, although this setup exhibited a much lower maximum efficiency.

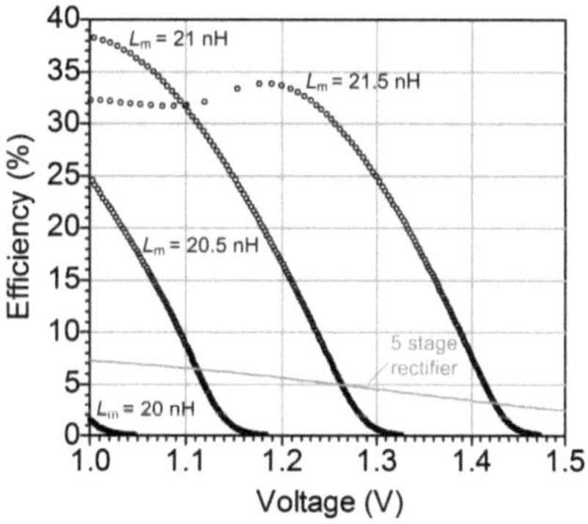

Fig. 6.17 Efficiency of a single-stage rectifier with $R_S = 50\ \Omega$ and an *LC* matching network ($C_m = 0.5$ pF and different values of L_m). The results obtained using a rectifier with $N = 5$ and $R_S = 300\ \Omega$ (*grey curve*) are plotted for comparison. $P_{AV} = -10$ dBm

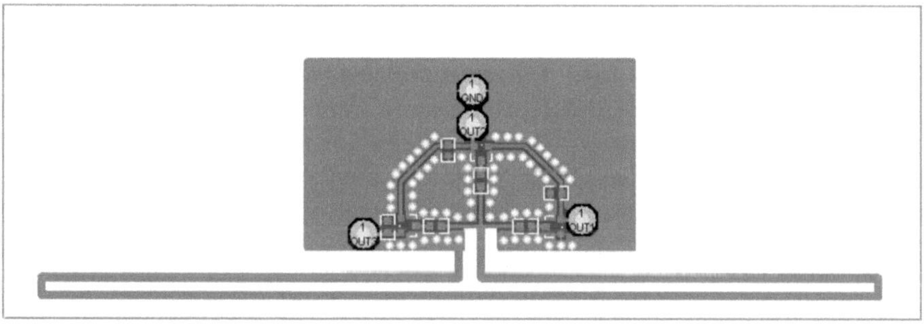

Fig. 6.18 Layout of the rectifier with the folded dipole attached

6.3 Experimental Results

We designed an antenna and rectifier to be connected to a storage unit comprising two NiMH batteries in series (V_{OUT} of 2.4 V to 3 V). As observed in Fig. 6.14, a rectifier with three stages offers good efficiency in this voltage range at $P_{AV} =$ 0 dBm. A PCB with a folded dipole antenna and a three-stage rectifier with shunt inductive matching network was fabricated. The layout of the circuit with the folded dipole ($\sim 300\ \Omega$ radiation antenna) and the rectifier is illustrated in Fig. 6.18. We simulated the folded dipole with Momentum (ADS) software to find the optimal dimensions of the antenna, finally selecting a length for the dipoles of 112 mm, with a separation between them of 1.22 mm, and a width of the traces of 1.01 mm.

Radiofrequency circuit design requires special care in the layout. To reduce the parasitic capacitance added by the layout, the ground plane must be well distributed over the entire PCB surface. Thus, several vias were used to connect the top and bottom sides of the ground plane. No simulations of this layout were performed; therefore, the added stray capacitances were not characterized.

We used a storage unit comprising two V20HR NiMH batteries (Varta) connected in series. Each battery has a capacitance of 20 mAh and a series resistance of 0.5 Ω. During the tests the battery voltage was 2.66 V. The practical value of L_{shunt} was found by trial and error using values for commercially available models. The best results at this voltage for $P_{AV} = 0$ dBm were obtained with $L_{shunt} = 4.7$ nH. This value was smaller than the values for $N = 3$ (see Table 6.2). This may result from the added stray capacitance, which was not considered during the simulations. Furthermore, the selected value of L_{shunt} optimized the efficiency at $V_{OUT} = 2.66$ V, but did not necessarily provide the best efficiency across the whole expected range of voltages of the storage unit.

6.3.1 Measurement Setup

The measurement setup comprised a transmitter and a receiver (Fig. 6.19). The transmitter was composed of an SMIQ 06B signal generator, a ZHL-42 power amplifier and a 50 Ω Kathrein monopole antenna (2 dBi gain). The power amplifier was used to achieve 0.5 W EIRP. The antenna had a bandwidth between 870 MHz and 960 MHz; therefore, 870 MHz, the nearest frequency to the ISM 868 MHz band, was chosen for operation. The signal generator was connected to the power amplifier with a 0.5 m cable. A 1 m cable was used to connect the power amplifier to the antenna. Cable losses were measured at the operating frequency for input powers between −20 dBm and 27 dBm. For this power range, the 1 m cable had 0.6 dB loss, whereas the 0.5 m cable had a 0.5 dB loss. The gain of the power amplifier was also measured and was found to be 37 dB at 15 V power supply and the operating frequency (870 MHz). The transmitted power (P_t) was varied between 27 dBm (0.5 W, maximum permitted) to 17 dBm. This was achieved by varying the output power of the signal generator (P_{gen}), using values ranging from −10.9 dBm to −20.9 dBm (in 1 dB steps). This power variation also enabled emulation of the case in which the transmitted power is fixed and the distance from the receiver to the transmitter is changed.

The receiver, which was previously designed, comprised the PCB with the folded dipole antenna, the three-stage rectifier and the battery. Output current and voltage were simultaneously measured with a PREMA 5017 Digital Multimeter. To avoid reflections, several radiation-absorbing panels were positioned around the transmitter. Due to the lack of an anechoic chamber, the distance between the two parts was kept short (35 cm), but still large enough to be in the far field region (> 20 cm). The received RF power was measured with a Kathrein monopole antenna and a FSIQ7 Vector Signal Analyzer (Rohde & Schwarz). A 1 meter cable was used to interconnect the antenna and the signal analyzer.

6.3.2 Results

Table 6.4 shows P_{gen}, P_t, the theoretical ($P_{r,t}$) and measured ($P_{r,m}$) received powers, and the equivalent distances ($d_{0.5}$ and d_4, respectively). The parameter $P_{r,t}$ was computed by subtracting P_t from the free-space path loss (22.12 dB, corresponding

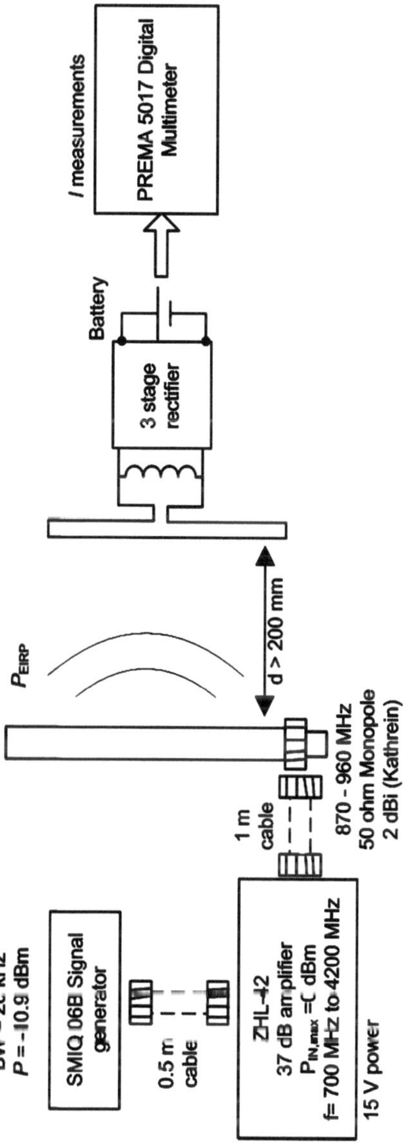

Fig. 6.19 Measurement setup

Table 6.4 Emitted and received power values and equivalent distances

P_{gen} (dBm)	P_t (dBm)	$P_{r,t}$ (dBm)	$P_{r,m}$ (dBm and mW)		$d_{0.5}$ (m)	d_4 (m)
−10.9	27	4.88	4.6	2.88	0.36	1.02
−11.9	26	3.88	3.5	2.23	0.41	1.16
−12.9	25	2.88	2.5	1.77	0.46	1.3
−13.9	24	1.88	1.5	1.41	0.51	1.46
−14.9	23	0.88	0.4	1.09	0.58	1.66
−15.9	22	−0.11	−0.5	0.89	0.64	1.84
−16.9	21	−1.11	−1.4	0.72	0.72	2.04
−17.9	20	−2.11	−2.4	0.57	0.8	2.29
−18.9	19	−3.11	−3.4	0.45	0.9	2.57
−19.9	18	−4.11	−4.4	0.36	1.02	2.88
−20.9	17	−5.11	−5.4	0.28	1.14	3.23

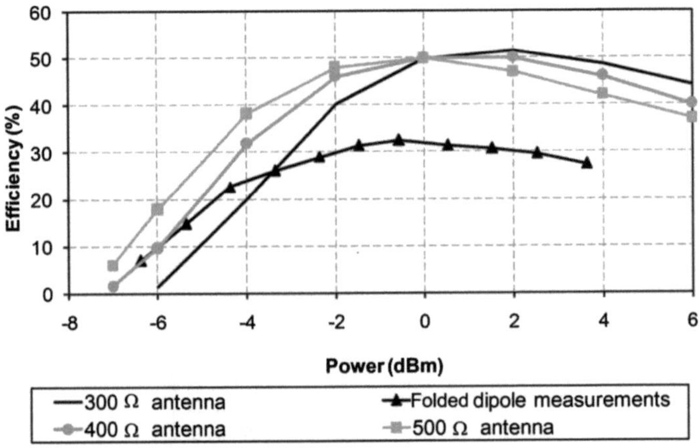

Fig. 6.20 Efficiency plotted against P_{AV} for a folded dipole connected to a three-stage rectifier. The output was connected to a battery (voltage = 2.66 V). Measurements were compared with simulations run at three values of R_S (300 Ω, 400 Ω and 500 Ω). For all simulations, L_{shunt} was 10.5 nH

to 35 cm). The parameters $d_{0.5}$ and d_4 refer to the equivalent distances at which the receiver should be placed from the transmitter in order to achieve the corresponding $P_{r,t}$ when transmitting at 0.5 W and 4 W EIRP, respectively.

Figure 6.20 shows the computed efficiency of the receiver when the transmitted power and, subsequently, the received power (Fig. 6.19), were varied, and compares it against simulations for different values of R_S (300 Ω, 400 Ω and 500 Ω). The efficiency was computed by using (6.11). P_{AV} was computed from the measured power and the gain of the antenna. The battery connected at the output presented a voltage of 2.66 V; this value was also used in the simulations. L_{shunt} was fixed at 10.5 nH in all simulations. This value maximized efficiency at 0 dBm (see Table 6.2). The tested folded dipole performed better at low power than did the 300 Ω simulated antenna. This could be due to the antenna's impedance, which is slightly higher. The shape of the efficiency curve agreed better with the simulations in which R_S was 400 Ω or 500 Ω than when R_S was 300 Ω. In fact, the antenna impedance is $300\ \Omega \cdot (1.469 + 0.332j)$. Thus, the real part of the antenna impedance is roughly 440 Ω. Nonetheless, the experimental efficiency was significantly lower than the simulation data for $P_{AV} > -4$ dBm. Possible explanations for this include a deficient antenna implementation (the final parameter values of the antenna have not been measured) and an improper value of L_{shunt}. Nonetheless, further work is required to improve the overall design and to assess the efficiency at different output voltages.

6.4 Conclusions

This chapter begins with a background on RF energy harvesting. All the blocks of an RF harvester are presented, including the antenna, the rectifier, and the impedance matching between them. Related works that deal with the different blocks as well as with full autonomous sensors are also discussed.

We performed simulations at several received powers at the antenna ($P_{AV} = -10$ dBm, 0 dBm and 10 dBm) and considering different rectifier circuit and matching impedance options. A rectifier multiplier with a variable number of stages (one to five) and a single diode rectifier was considered, as were shunt inductor and *LC* matching impedance networks. For the antenna, a voltage source with an output impedance (50 Ω and 300 Ω) was used.

We ran one set of simulations with the rectifier multiplier, shunt inductor matching, and 300 Ω antenna. Circuit efficiency was observed to slightly depend on the number of stages used for the rectifier multiplier, but varied markedly with received power: it was roughly 10%, 50% and 80% at power levels of −10 dBm, 0 dBm, and 10 dBm, respectively. The efficiency was plotted against the output voltage of the rectifier at the aforementioned power levels and for a different number of rectifier stages (one to five). The output voltage that marks maximum efficiency (V_{MPP}) increases with increasing power level and number of stages. For $P_{AV} = 10$ dBm, a single-stage rectifier was good enough, as it led to a V_{MPP} of roughly 4 V. Thus, it could directly recharge one Li-ion battery or three series-connected NiMH batteries with good efficiency. A higher number of stages led to higher values of V_{MPP}. For $P_{AV} = 0$ dBm, V_{MPP} increased in steps of 1 V from 1 V to 5 V when the number of stages of the rectifier increased from 1 to 5. Thus, a four-stage rectifier can be used to recharge a single Li-ion battery or three series-connected NiMH batteries. Rectifiers with one or two stages could be used to recharge storage units of one NiMH or two series-connected NiMH batteries, respectively. For $P_{AV} = -10$ dBm, V_{MPP} was always lower than 1 V. Consequently, a higher number of stages should be used to recharge at least 1 NiMH battery with good efficiency. In all cases, an MPPT circuit placed between the output of the rectifier and the storage unit enabled the system to work at its maximum power point. However, due to the power waste of the MPPT circuit, this solution is only feasible if it offers a net power gain.

For $P_{AV} = 10$ dBm, a single diode rectifier achieved an efficiency of less than 65%, which is worse than that achieved with the rectifier multiplier. A single-stage rectifier multiplier with *LC* matching network and antenna impedance of 50 Ω achieved, at $P_{AV} = 0$ dBm, an efficiency greater than 60% and a V_{MPP} of roughly 2 V, and at $P_{AV} = -10$ dBm, an efficiency of less than 40% and a V_{MPP} of approximately 1 V. However, good *LC* matching networks can be hard to implement, since they are highly sensitive to variations in the components and to the parasitic resistances of the matching elements, which lower the effective Q of the system.

Finally, we ran experimental tests with a folded dipole antenna (about 300 Ω impedance) and a three-stage rectifier multiplier, implemented with off-the-shelf components. Radiofrequency power was deliberately radiated and the received power levels at the antenna ranged from approximately −5 dBm to 5 dBm. A storage unit comprising two serial-connected NiMH batteries (total voltage: 2.66 V) was used. Maximum efficiency was greater than 30% for power levels near 0 dBm, which is well below the simulation result of roughly 50%.

References

1. K. Finkenzeller, *RFID Handbook Fundamentals and Applications in Contactless Smart Cards and Identification*, 2nd edn. RFID Handbook. Fundamentals and Applications in Contactless Smart Cards and Identification (John Wiley and Sons, West Sussex, 2003)
2. J.-P. Curty, M. Declercq, C. Dehollain, N. Joehl, *Design and Optimization of Passive UHF RFID Systems*, 1st edn. (Springer Science+Business Media, New York, 2007)
3. M. Minhong, M.H. Mickle, C. Capelli, H. Swift, RF energy harvesting with multiple antennas in the same space. IEEE Antennas and Propagation Magazine **47**, 100–106 (2005)
4. M. Ali, G. Yang, R. Dougal, A new circularly polarized rectenna for wireless power transmission and data communication. IEEE Antennas and Wireless Propagation Letters **4**, 205–208 (2005)
5. J.O. McSpadden, T. Yoo, K. Chang, Theoretical and experimental investigation of a rectenna element for microwave power transmission. IEEE Trans. Microwave Theory and Techniques **40**, 2359–2366 (1992)
6. S.A. Bhalerao, A.V. Chaudhary, R.B. Deshmukh, R.M. Patrikar, Powering wireless sensor nodes using ambient RF energy, in *Proceedings of IEEE International Conference on Systems, Man and Cybernetics, SMC* (2006), pp. 2695–2700
7. D.M. Dobkin, *The RF in RFID. Passive UHF RFID in Practice* (Elsevier, Burlington, 2008)
8. T. Le, K. Mayaram, T. Fiez, Efficient far-field radio frequency energy harvesting for passively powered sensor networks. IEEE Journal of Solid-State Circuits **43**, 1287–1302 (2008)
9. J.A. Hagerty, F.B. Helmbrecht, W.H. McCalpin, R. Zane, Z.B. Popovic, Recycling ambient microwave energy with broad-band rectenna arrays. IEEE Trans. on Microwave Theory and Techniques **52**, 1014–1024 (2004)
10. W.C. Brown, An experimental low power density rectenna, in *Proceedings of IEEE MTT-S International Microwave Symposium Digest*, vol. 1 (1991), pp. 197–200
11. R.E. Barnett, L. Jin, S. Lazar, A RF to DC voltage conversion model for multi-stage rectifiers in UHF RFID transponders. IEEE Journal of Solid-State Circuits **44**, 354–370 (2009)
12. J.P. Curty, N. Joehl, C. Dehollain, M.J. Declercq, Remotely powered addressable UHF RFID integrated system. IEEE Journal of Solid-State Circuits **40**, 2193–2202 (2005)

13. F. de Dieuleveult, Electrónica aplicada a las altas frecuencias (Paraninfo, 2001)
14. S. Young-Ho, C. Kai, A high-efficiency dual-frequency rectenna for 2.45- and 5.8-GHz wireless power transmission. IEEE Trans. Microwave Theory and Techniques **50**, 1784–1789 (2002)
15. M. Ghovanloo, K. Najafi, Fully integrated wideband high-current rectifiers for inductively powered devices. IEEE Journal of Solid-State Circuits **39**, 1976–1984 (2004)
16. J.P. Curty, N. Joehl, F. Krummenacher, C. Dehollain, M.J. Declercq, A model for u-power rectifier analysis and design. IEEE Trans. Circuits and Systems I: Regular Papers **52**, 2771–2779 (2005)
17. U. Karthaus, M. Fischer, Fully integrated passive UHF RFID transponder IC with 16.7 uW minimum RF input power. IEEE Journal of Solid-State Circuits **38**, 1602–1608 (2003)
18. T. Umeda, H. Yoshida, S. Sekine, Y. Fujita, T. Suzuki, S. Otaka, A 950-MHz rectifier circuit for sensor network tags with 10-m distance. IEEE Journal of Solid-State Circuits **41**, 35–41 (2006)
19. P. Spies, F. Förster, L. Mateu, M. Pollak, Power management in energy harvesting power supplies. Presented at 1st International Workshop on Power Supply on Chip (PwrSoC), Cork, Ireland, September 22.
20. T. Paing, J. Shin, R. Zane, Z. Popovic, Resistor emulation approach to low-power rf energy harvesting. IEEE Trans. Power Electronics **23**, 1494–1501 (2008)
21. J.R. Smith, A.P. Sample, P.S. Powledge, S. Roy, A. Mamishev, A wirelessly-powered platform for sensing and computation. Available on: http://web.media.mit.edu/~jrs/WISP-Ubicomp-06.pdf. Accessed June 2007
22. F. Kocer, M.P. Flynn, An RF-powered, wireless CMOS temperature sensor. IEEE Sensors Journal **6**, 557–564 (2006)
23. T. Paing, J. Morroni, A. Dolgov, J. Shin, J. Brannan, R. Zane, Z. Popovic, Wirelessly powered wireless sensor platform, in *Proceedings of European Microwave Conference* (2007), pp. 999–1002
24. P. Singh, W. Xiquan, R. Lafollette, D. Reisner, RF-recharged microbattery for powering miniature sensors, in *Proceedings of IEEE Sensors*, vol. 1 (2004), pp. 349–352
25. P. Mars, C. Greene, RF energy harvesting and battery-free wireless sensors. Available on: http://www.cap-xx.com/resources/docs/CAP-XX&Powercast_RF_Energy.pdf. Accessed January 2010
26. C.A. Balanis, *Antenna Theory: Analysis and Design*, 3rd edn. (2005)
27. R.W. Erickson, D. Maksimovic, *Fundamentals of Power Electronics*, 2nd edn. (Kluwer Academic, Norwell, 2001)

MIX
Papier aus verantwortungsvollen Quellen
Paper from responsible sources
FSC® C105338

If you have any concerns about our products,
you can contact us on
ProductSafety@springernature.com

In case Publisher is established outside the EU,
the EU authorized representative is:
Springer Nature Customer Service Center GmbH
Europaplatz 3, 69115 Heidelberg, Germany

Printed by Libri Plureos GmbH
in Hamburg, Germany